マリタイムカレッジシリーズ

1・2級海技士
はじめての英語指南書

商船高専キャリア教育研究会 編

専門知識につなげよう

KAIBUNDO

◎執筆者一覧
　CHAPTER 1, 3
　　長山昌子（富山高等専門学校）
　　勝島隆史（富山高等専門学校）
　　岸　拓真（広島商船高等専門学校）
　CHAPTER 2, 4
　　長山昌子
　　経田僚昭（富山高等専門学校）
　　篠島司郎（富山高等専門学校）
　コラム
　　橋爪仙彦（鳥羽商船高等専門学校）
　　野口　隆（弓削商船高等専門学校）
　イラスト制作
　　前畑航平（大島商船高等専門学校）

◎編集幹事
　勝島隆史
　経田僚昭

はじめに

　本書は，2級および1級海技士国家試験に出題される英語の問題の内容を，学習者のみなさんが正しく理解する力をつけて，学生から外航船員・内航船員へ成長するためのかけはしとなればと思い，制作しました。

　国家試験には，外航船員の業務で使用される書類や国際条約が，多く出題されています。なかには，授業で取り扱わないような高い専門性が要求される内容も多くあります。海技士国家試験に合格するには，学校の授業で学んだそれぞれの知識同士を結びつけて，船舶運航について総合的に理解する必要があります。それには，単に英文を理解して和訳するだけでなく，英語の基礎的な知識や，船舶運航に関する幅広い専門用語や専門知識を身につける必要があります。そして，船員として安全かつ的確に船舶運航を行うには，学問としての知識だけではなく，現場ですぐさま応用できる実践的で総合的な知識が求められています。それには，学校で学んだ知識を基礎にしながらも，現場での経験を通じて自分のなかに生きた知識を構築する必要があります。本書は，この大きな目標の第一歩として，学習者のみなさんが自分一人の力でも，英文法や専門用語，専門知識を無理なく正しく理解し，効果的に力をつけられるように工夫しました。

　本書のCHAPTER 1と2では，実際に出題された問題を用いて，英文法や語彙，専門用語の解説をしました。航海20問，機関20問，計40問の問題が収められています。難解な専門用語の意味がスムーズに理解でき，長文読解のコツがわかり，各自で学習を進められるように3つのステップを用意しました。STEP 1では，自分で，英文の仕組みを考えて，意味のまとまりをつかんでください。STEP 2では，語彙の解説を読みながら，STEP 1で理解した英文の仕組みや意味のまとまりを確認してください。そして，部分訳や全文訳を書き込んでください。STEP 3では，音読しながら，スラッシュや主語・述語動詞を意識してください。読めるようになると，語の定着も早くなります。学習の進度を確かめるために，各ステップが終了したら，チェックボックスに印をつけてください。Unit 毎に，専門用語の語彙を増やすために単語リストを掲げ，学習者が意味を書き込むようにしました。また，専門知識を深めるとともに正しく理解するために，重要な内容や専門性の高い内容については，Unit の最後に，図解説明を掲載しました。日本語でしっかり内容を把握すれば，専門知識がわかり，英文を深く理解するのに有効です。

　CHAPTER 3と4では，高度な専門性を身につけるために，外航船員の業務で使用される書類や国際条約から重要な文章を選定し，掲載しました。どちらもCHAPTER 1と2と同様に主語や述語動詞の区別，スラッシュが記載されています。はじめて見る文書が多いと思いますが，これらをしっかりマスターすれば，これから受験する海技士試験，その後のさまざまな状況のもとで英語で書かれた文書を理解しなければならないときに，あせることなく品詞，ときにはスラッシュを意識し，その文書中のキーワードを調べることで，内容が把握できると思います。本書で学んだ方法がその一助になれば幸いです。そのために，CHAPTER 3と4ではあえて解説や和訳を付けず，単語の意味の記載も最小限としました。

また，航海士・機関士は同じ船で生活を共にし，互いに協力しながら船舶運航に取り組んでいます。互いの仕事内容を理解することが，円滑な船舶運航につながります。航海士（航海の勉強をしている学生）の方はCHAPTER 4に，機関士（機関の勉強をしている学生）の方はCHAPTER 3に，ぜひともチャレンジしてください。各Unitのタイトルを見て興味のある内容を読んでみてください。どちらも同様に品詞の区別・スラッシュが記載されていますので，わからない語を辞書で調べながら取り組めば，きっと読み進めることができると思います。学習者が，高度な専門性を要求される英文に挑戦することを期待しています。図解説明があるので，自ら理解するのに役に立つはずです。

　また，外航船は多国籍の船員で運航されており，船内でスムーズにコミュニケーションをとるには，正しい英語の発音を身につけることが基本になります。本書に出てくる英文や単語については，ネイティブスピーカによるリスニング音声を収録し，ホームページに掲載しました(http://www.kaibundo.jp/sinan.htm)。各自のスマートフォンやタブレット，パソコンにダウンロードし，いつでもどこでも繰り返し聞くことにより，正しい英語の発音が身につけられるようにしました。右のコードでアクセスできます。

　末筆になりますが，大学間連携共同教育推進事業「海事分野における高専・産業界連携による人材育成システムの開発」のプロジェクトの一環として，本書は制作されました。本書の執筆・刊行にあたって，国際船員労務協会，一般社団法人日本船長協会，一般社団法人日本船舶機関士協会から，CHAPTER 3および4の英文の提供などの全面的な支援を受けました。また，海文堂出版の岩本登志雄氏からは貴重なアドバイスとご助力をいただきました。本書の執筆・刊行にあたってお世話になったみなさまに，厚くお礼申し上げます。

　2級および1級海技士国家試験に挑戦される学習者のみなさんが，本書を十分に活用して英語の基礎知識や船舶運航の専門知識を身につけ，外航船員・内航船員として活躍されることを願っています。

2015年8月
商船高専キャリア教育研究会
本書執筆代表者
勝島隆史
経田僚昭

目次

執筆者一覧……2
はじめに……3
本書における英単語の登場回数ランキング
　　　　……6

CHAPTER 1　海技士問題　航海……7
- Unit 1　航海当直1……8
- Unit 2　航海当直2……12
- Unit 3　航行の安全 自動操舵の使用……16
- Unit 4　海上労働 年少船員……20
- Unit 5　船員の訓練1……24
- Unit 6　船員の訓練2……28
- Unit 7　船員の資格証明……32
- Unit 8　沿岸航海に従事する船員の要件……36
- Unit 9　遭難信号……40
- Unit 10　海難救助と責任……44
- Unit 11　海難救助捜索1……48
- Unit 12　海難救助捜索2……52
- Unit 13　救命いかだ……56
- Unit 14　積荷に対する責任……60
- Unit 15　海難救助と積荷の避難……64
- Unit 16　積荷の注意事項……68
- Unit 17　カーゴタンクの原油洗浄……72
- Unit 18　バラスト水……76
- Unit 19　船員の保護設備……80
- Unit 20　条約の適用……84

CHAPTER 2　海技士問題　機関……89
- Unit 21　主機空気制御……90
- Unit 22　カム・カム軸……94
- Unit 23　排気弁……98
- Unit 24　ターボチャージャー……102
- Unit 25　給水処理……106
- Unit 26　ボイラ給水ポンプ……110
- Unit 27　軸心調整……114
- Unit 28　プロペラ軸……118
- Unit 29　冷凍機……122
- Unit 30　LO清浄系統……124
- Unit 31　メカニカルシール……130
- Unit 32　発電機……134
- Unit 33　単相と三相……140
- Unit 34　サーキットブレーカー……144
- Unit 35　ダイオード……148
- Unit 36　排ガス規制……152
- Unit 37　主機の据付け……156
- Unit 38　高温腐食……162
- Unit 39　ピストン……166
- Unit 40　スラッジ処理……170

CHAPTER 3　専門英語　航海……175
- Unit 41　操縦性能の基礎……178
- Unit 42　当直に関する基準……180
- Unit 43　航行の安全……184
- Unit 44　捜索パターン……186
- Unit 45　船位通報制度……188
- Unit 46　運航に伴う油の排出規制……190
- Unit 47　調整の許容範囲……194

CHAPTER 4　専門英語　機関……193
- Unit 48　調整の許容範囲……196
- Unit 49　非破壊試験……196
- Unit 50　ダイオード……198
- Unit 51　ターボチャージャーのサージング……200
- Unit 52　ピストンクラウンの損傷……202
　　　　（バナジウムアタック）
- Unit 53　ピストンクラウンの焼損……204
- Unit 54　船尾管軸受の損傷……206

- コラム 1　述語動詞の見つけ方……88
- コラム 2　助動詞……121
- コラム 3　前置詞1……139
- コラム 4　前置詞2 ofの訳し方……151
- コラム 5　前置詞3 熟語について……161
- コラム 6　接続詞andの用法について……173
- コラム 7　thatの用法について……174
- コラム 8　接続詞のthatと関係代名詞のthatの見分け方……183
- コラム 9　関係代名詞thatの使い方……189
- コラム 10　so～thatとso thatの使い方……192

本書における英単語の登場回数ランキング

順位	回数	単語
1	924	the
2	508	be
3	461	of
4	348	and
5	284	to
6	279	a
7	217	in
8	122	or
9	111	ship
10	103	for
11	87	that
12	85	as*
13	84	**shall**
14	82	with
15	80	on
16	78	by
17	72	it
18	69	at
19	63	this
20	59	when
⋮		
22	51	from
⋮		
27	41	**should**
⋮		
39	29	into
⋮		
53	24	**may**
⋮		
63	20	**will**
⋮		
79	17	than*
⋮		
91	15	during
91	15	over*
91	15	through
91	15	under

順位	回数	単語
105	14	between
⋮		
119	13	**can**
⋮		
146	11	**must**
⋮		
165	10	before*
165	10	**would**
⋮		
189	9	above
⋮		
215	8	within
⋮		
248	7	near
⋮		
359	5	without
⋮		
440	4	about*
440	4	against
440	4	along
440	4	around*
440	4	except
440	4	inside*
⋮		
566	3	**could**
566	3	**might**
566	3	toward
566	3	towards
⋮		
759	2	onto
759	2	upon
⋮		
1070	1	among
1070	1	beyond
1070	1	unlike

ゴシック体は助動詞

網掛けは前置詞

＊は品詞が異なっても同じ単語としてカウントしている

海技士問題 航海

Unit 1　航海当直 1

Step Check Box □ □ □

Step 1　下記の英文を読み，次の3つの活動をしなさい。終了後 Step Check Box □ をぬりなさい。

(1) 動詞（述語動詞）を○で囲む。動詞の意味を考えて，動詞のあとに続く内容を推測する。
(2) 主語（主部）に下線を引く。
(3) スラッシュを入れる。スラッシュを入れる場所は主に下記の3つ。
- コンマ(,)やピリオド(.)でスラッシュ(/)を入れる。音読するとき，スラッシュで息つぎする。
- 接続詞の前にスラッシュを入れる。［接続詞＋主語＋動詞］は文としてまとまった意味を表す。
- 原則として前置詞の前にスラッシュを入れる。（前置詞＋名詞）は，最小の意味のまとまりを表す。

1) Officers of the navigational watch shall know the handling characteristics of their ship, including its stopping distances, and should appreciate that other ships may have different handling characteristics.

2) It is of special importance that at all times the officer in charge of the navigational watch ensures that a proper look-out is maintained.

3) In a ship with a separate chart room the officer in charge of the navigational watch may visit the chart room, when essential, for a short period for the necessary performance of navigational duties, but shall first ensure that it is safe to do so and that proper look-out is maintained.

4) Operational tests of shipboard navigational equipment shall be carried out at sea as frequently as practicable and as circumstances permit, in particular before hazardous conditions affecting navigation are expected.

5) Whenever appropriate, these tests shall be recorded.

6) Such tests shall also be carried out prior to port arrival and departure.

出典：2級海技士（航海）平成21年4月定期試験問題
STCW条約　STCWコード, Part A Chapter 8, Part 4-1

Step 2　網掛けは主語，ゴシック体は動詞（述語動詞），＊語注・文法的留意事項，①②③は訳順番号

(1) 単語や熟語の説明を読みながら，部分訳を書きなさい。わからない単語の意味は調べなさい。
(2) 部分訳が終わったら，①②③の訳順番号を参考に，全文和訳を書きなさい。
　　全文和訳が終わったら，Step Check Box □ をぬりなさい。

1) ② ① ⑤ ④ ③
Officers of the navigational watch shall know the handling characteristics of their ship,/ including its stopping distances,/

* navigational watch：航海当直
* shall：〔助動詞〕主に法律文で使われるときは，〜すべしと訳す。
* the handling characteristics：操縦性能
* including：〔前置詞〕含めて
* stopping distance：停止距離
* ③の前置詞以下は，④にかかる。
* この第1文を読むと，「航海当直者が知っておくべき知識や任務」が，問われる内容だと推測できる。

①②部分訳_____　③④部分訳_____

⑦ ⑥
and (officers) should appreciate/ [that other ships may have different handling characteristics.]/

* appreciate：〔他動詞〕= understand
* 主語は，②と同じなので，ここでは省略されている。
* that：〔接続詞〕that 以下には主語 S ＋動詞 V が必要である。appreciate の目的語になっている。

⑥部分訳_____ということを

全文和訳_____

2) It **is** of special importance/[that at all times the officer in charge of the navigational watch **ensures**/
that a proper look-out **is maintained**.]/
(circled numbers: ⑥ It ⑤ is ① that at all times ② the officer in charge of the navigational watch ④ ensures ③ that a proper look-out is maintained)

* it：仮主語なので，「それは」と訳さない。that 以下が真主語なので，that 以下全文を「～ということは」と訳す。
  that 以下とは，ここでは [that at all times the officer … is maintained] の全文である。
* of importance：of ＋抽象名詞は形容詞として働くので，important と同じ意味
* at all times：＝ always
* in charge of：～に責任がある   * in charge of the navigational watch は，the officer にかかる。
* ensure：〔他動詞〕～を確認する   * that a proper look-out is maintained は，確認する (ensure) 内容を示している。
* a proper look-out：適切な見張り
* is maintained：〔受動態〕維持されている

①②部分訳 _____ が，いつも

③④部分訳 _____ ということを確認する

全文和訳 _____

3) In a ship/ with a separate chart room/ the officer in charge of the navigational watch
(② In a ship ① with a separate chart room ④ the officer in charge of ③ the navigational watch)

* with a separate chart room：(船橋と) 独立した海図室がついている (～がある)
* separate：〔形容詞〕独立した，別個の

①②部分訳 _____   ③④部分訳 _____

may visit the chart room,/ [when (**it is**) essential,/ for a short period/
for the necessary performance of navigational duties,]/
(⑧ may visit ⑦ when it is essential ⑥ for a short period ⑤ for the necessary performance of navigational duties)

* may：〔助動詞〕してもよい   * for the necessary performance：必要な仕事のために
* it is essential：(it is) が省略されている。 * navigational duties：航海上の職務
* for a short period：短時間   * 前置詞 (for, in, on, including など) の後には名詞が続く

⑤部分訳 _____   ⑥⑦部分訳 _____

but (the officer) **shall** first **ensure**/ [that it **is** safe to do so]/
(⑨ but the officer ⑫ shall first ensure ⑩ that it is safe to do so)

* the officer：この英文の動詞は，⑧ may visit と⑫ shall first ensure で，主語は同じ③④である。
  ここでは主語が省略されている。
* shall first ensure：[that 以下] を，確認するべきである
* it is safe to do so：it は仮主語，to do so は真主語。「そうすることが safe である」と訳す。
* 動詞⑫ ensure の具体的な内容は，⑩と⑪である。

⑩部分訳　しかし，士官は _____ ということと→⑪につづく

and (the officer **ensure**)/ [that proper look-out **is maintained**.]/
(⑪ that proper look-out is maintained)

* and：〔接続詞〕は [that it is safe to do so] と [that proper look-out is maintained] をつなぐ。
* is maintained：(適切な見張りが) 維持されている

⑪部分訳 _____ ということを最初に確認すべきである

全文和訳 _____

4) Operational tests of shipboard navigational equipment **shall be carried out** at sea/

- ＊ operational test：動作試験
- ＊ shipboard：〔形容詞〕船上での
- ＊ equipment：装置，装備
- ＊ operational：〔形容詞〕
- ＊ operation：〔名詞〕
- ＊ shall be carried out：〔受動態〕実施されるべきである
- ＊ at sea：海で

①②部分訳 _____

as frequently as practicable/ and (as frequently) as **circumstances permit**,/

- ＊ as 副詞 as practicable：基本形は as soon as possible（できるだけ早く）である。
- ＊ frequently：〔副詞〕頻繁に
- ＊ practicable：〔形容詞〕実行可能な ＊ practice：〔他動詞〕
- ＊ as frequently as practicable：実行可能なだけ頻繁に
- ＊ circumstance：環境
- ＊ permit：許す
- ＊ (as frequently) as circumstances permit：as frequently が省略されている。

⑧⑨部分訳 _____

in particular/ [before hazardous conditions affecting navigation **are expected**.]/

- ＊ in particular：とくに
- ＊ hazardous：〔形容詞〕危険な
- ＊ hazard：〔名詞〕＝ danger
- ＊ condition：〔名詞〕状況
- ＊ affect：〔他動詞〕（直接的に）に影響する
- ＊ navigation：〔名詞〕航海
- ＊ affecting：〔現在分詞形〕～している　（ここでは直前の名詞 hazardous conditions にかかる）
- ＊ expect：〔他動詞〕～を予想(予期)する
- ＊ before：〔接続詞〕なので，before の後には主語Sと動詞Vがくる。

④⑤⑥部分訳 _____

全文和訳
_____

5) [Whenever (they **are**) appropriate,]/ these tests **shall be recorded**./

- ＊ whenever：〔接続詞〕～のときはいつでも
- ＊ (they are) が省略されている。they ＝ these tests
- ＊ appropriate：〔形容詞〕適切な
- ＊ shall be recorded：〔受動態〕記録されるべきである

全文和訳
_____

6) Such tests **shall** also **be carried out**/ prior to port arrival and (port) departure./

- ＊ such tests：そのようなテスト（operational tests のこと）
- ＊ prior to：〔前置詞〕～より前に（＝ before）
- ＊ port arrival：入港
- ＊ (port) departure：出港

全文和訳
_____

Step 3　網掛けは主語，**ゴシック体**は動詞（述語動詞）

(1) 音声を聞きながら，音読する。音読が終わったら，Step Check Box □をぬりなさい。
(2) 左から順に，スラッシュ毎の意味を考える。

Officers of the navigational watch **shall know** the handling characteristics of their ship,/ including its stopping distances,/ and **should appreciate**/ that other ships **may have** different handling characteristics./ It **is** of special importance/ that at all times the officer in charge of the navigational watch **ensures**/ that a proper look-out **is maintained**./ In a ship/ with a separate chart room/ the officer in charge of the navigational watch **may visit** the chart room,/ when essential,/ for a short period/ for the necessary performance of navigational duties,/ but **shall** first **ensure**/ that it **is** safe to do so/ and that proper look-out **is maintained**./ Operational tests of shipboard navigational equipment **shall be carried out** at sea/ as frequently as practicable/ and as circumstances **permit**,/ in particular/ before hazardous conditions affecting navigation **are expected**./ Whenever appropriate,/ these tests **shall be recorded**./ Such tests **shall** also **be carried out**/ prior to port arrival and departure./

【専門英語単語リスト】書いておぼえよう

発音したり，スペル練習したり，意味を調べたりしたらチェック欄の○をぬりなさい。
辞書を引いて単語の意味が多くある場合，英語の文脈から判断し，適切な意味を選びなさい。

英文	単語	品詞	意味を調べて書きなさい	発音しながらスペル練習しなさい	チェック
1)	navigational watch	熟語			○○○
1)	distance	名詞			○○○
1)	appreciate	他動詞			○○○
1)	characteristic	名詞			○○○
2)	look-out	名詞			○○○
2)	maintain	他動詞			○○○
3)	chart room	熟語			○○○
3)	performance	名詞			○○○
3)	duty	名詞			○○○
4)	equipment	名詞			○○○
4)	carry out	他動詞			○○○
4)	circumstance	名詞			○○○
4)	permit	他動詞			○○○
4)	affect	他動詞			○○○
4)	expect	他動詞			○○○

理解を深めるキーワード「航海当直基準」

　外航船員が遵守すべき重要な国際条約の一つに，船員の訓練や当直基準などを定めた STCW 条約（1978 年の船員の訓練及び資格証明並びに当直の基準に関する国際条約；The International Convention on Standards of Training, Certification and Watchkeeping for Seafarers, 1978）がある。この条約では，船員の質の向上により海上の安全を増進することを目的としている。日本は STCW 条約を批准しており，条約の規定内容が船員法や船舶職員及び小型船舶操縦者法などの国内法に反映されている。この章で取り扱った当直基準の他に，当直を実施するときの船員の休息時間や資格，航海計画に加えて，当直の際の見張りや引き継ぎ，当直体制，当直実施の遵守事項が定められている。これらの内容は，船舶の安全を確保するために，どのようにして当直を行うかを詳細に定めたものであることから，当直を担当する航海士・機関士は，これらの規定内容を熟知した上で当直に臨む必要がある。詳細については，Unit 43 に示す STCW 条約の当直基準を参照のこと。

Unit 2　航海当直 2

Step Check Box □ □ □

Step 1　下記の英文を読み，次の3つの活動をしなさい。終了後 Step Check Box □をぬりなさい。

(1) 動詞（述語動詞）を○で囲む。動詞の意味を考えて，動詞のあとに続く内容を推測する。
(2) 主語（主部）に下線を引く。
(3) スラッシュを入れる。スラッシュを入れる場所は主に下記の3つ。
- コンマ (,) やピリオド (.) でスラッシュ (/) を入れる。音読するときは，スラッシュで息つぎする。
- 接続詞の前にスラッシュを入れる。［接続詞＋主語＋動詞］は文としてまとまった意味を表す。
  ここでいう接続詞とは when, if, until, before, after など (and, but, or, so を含まない)。
- 原則として前置詞の前にスラッシュを入れる。（前置詞＋名詞）は，最小の意味のまとまりを表す。

1) The officer in charge of the navigational watch shall take frequent and accurate compass bearings of approaching ships as a means of early detection of risk of collision and bear in mind that such risk may sometimes exist even when an appreciable bearing change is evident, particularly when approaching a very large ship or a tow or when approaching a ship at close range.

2) The officer in charge of the navigational watch shall also take early and positive action in compliance with the applicable International Regulations for Preventing Collisions at Sea, 1972 and subsequently check that such action is having the desired effect.

3) In clear weather, whenever possible, the officer in charge of the navigational watch shall carry out radar practice.

出典：2級海技士（航海）平成24年7月定期試験問題
STCW 条約　STCW コード　A-Ⅷ 43, 44 からの抜粋

Step 2　網掛けは主語，**ゴシック体**は動詞（述語動詞），＊語注・文法的留意事項，①②③は訳順番号

(1) 単語や熟語の説明を読みながら，部分訳を書きなさい。わからない単語の意味は調べなさい。
(2) 部分訳が終わったら，①②③の訳順番号を参考に，全文和訳を書きなさい。
　　全文和訳が終了したら，Step Check Box □をぬりなさい。

1)　② ①　⑥　⑤
The officer in charge of the navigational watch **shall take** frequent and accurate compass bearings/

- ＊ in charge of：〔熟語〕〜の責任がある
- ＊ officer in charge of the navigational watch：当直航海士
- ＊ navigational：〔形容詞〕　　＊ navigate：〔他動詞〕　　＊ navigation：〔名詞〕
- ＊ frequent and accurate compass bearings：頻繁で正確な（④近づいてくる他船への）コンパス方位
- ＊ bearings：（ある目標に対する自船からの）方位，方向

①②④⑤⑥部分訳 _____

　　④　　　　　　　　　③　　　　　　　　　　　⑦　　⑯
of approaching ships/ as a means of early detection of risk of collision/ and (shall) **bear in mind**/

- ＊ approach：〔自動詞〕近づく
- ＊ approaching：〔現在分詞〕近づいてくる，ship を修飾する形容詞の働きをする。
- ＊ as a means of 〜：〔熟語〕〜の手段として
- ＊ detection：〔名詞〕発見，探知　　　　　　＊ detect：〔他動詞〕〜を探知する，発見する
- ＊ risk of collision：衝突の危険　　　　　　　＊ collide：〔自動詞〕〜と衝突する
- ＊ 接続詞⑦ and は，何と何をつなぐか？→動詞 shall take と動詞 (shall) bear in mind をつなぐ。
- ＊ bear in mind = remember, take into account
- ＊ bear in mind that such risk 〜：⑮［that such risk may sometimes exist］を覚えているべきである

③部分訳 _____　⑮⑯部分訳 _____

12

[that such risk ⑮**may** sometimes **exist**]/ [even when ⑧an appreciable bearing change **is** evident,]/

* such risk：そのような危険　　　　　　　　　　* exist：〔自動詞〕存在する　　　　　　　　* existence：〔名詞〕
* even when S + V…：〜のときでさえ
* appreciable：〔形容詞〕目に見えるほどの，識別できる程度の　　* appreciate：〔他動詞〕識別する，高く評価する
* evident：〔形容詞〕明白な = clearly seen or understood, obvious　　* evidence：〔名詞〕証拠

⑧部分訳 ＿＿＿＿＿＿＿＿＿＿＿＿＿＿＿＿＿＿＿＿＿＿＿＿＿＿＿＿のときでさえ

[⑨particularly when (is) ⑪**approaching** a very large ship ⑩or a tow/] ⑫or

* particularly when V + S：S と V の位置が倒置。「とくに〜のときに」の意味。
* tow：〔名詞〕引き船（tug boat）　　　　　　　* or：〔接続詞〕あるいは
* 接続詞 or は何と何をつなぐか？→［⑨⑩⑪］あるいは［⑬⑭］

⑨⑩⑪部分訳 ＿＿＿＿＿＿＿＿＿＿＿＿＿＿＿＿＿＿＿＿＿＿＿＿のとき，あるいは

[when (is) ⑭**approaching** ⑬a ship/ at close range.]/

* when V + S：S と V の位置が倒置になっている。　　* at close range：狭い領域で

⑬⑭部分訳 ＿＿＿＿＿＿＿＿＿＿＿＿＿＿＿＿＿＿＿＿＿＿＿＿のときに

全文和訳

* 主語の the navigational watch という単語から，当直中の航海士の仕事が問われていると推測し，動詞の後に続く英文を訳す。専門科目に関する幅広い知識があれば，英文理解に大変役に立つ。

2) ②The ①officer in charge of the navigational watch ⑥**shall** also ⑤**take** early and positive action/

* take action：〔熟語〕行動をとる　　　　　　　* early and positive：早期に積極的な

①②⑤⑥部分訳 ＿＿＿＿＿＿＿＿＿＿＿＿＿＿＿＿＿＿＿＿＿＿＿＿＿＿＿＿＿＿＿＿＿＿＿

in compliance with the ④applicable ③International Regulations for Preventing Collisions at Sea, 1972/

* in compliance with：〔熟語〕〜に従って（応じて）
* applicable：〔形容詞〕適用できる，実用的な　　　* apply：〔他動詞〕〜を適用する，利用する
* International Regulations for Preventing Collisions at Sea, 1972：COLREG 条約（1972 年の国際海上衝突予防規則に関する条約）
* in compliance with the applicable 〜：「〜の条約を適用して」と，わかりやすく訳してもよい。

③④部分訳 ＿＿＿＿＿＿＿＿＿＿＿＿＿＿＿＿＿＿＿＿＿＿＿＿＿＿＿＿＿＿＿＿＿＿＿＿＿＿

⑦and (shall) ⑧subsequently ⑩**check**/ [that ⑨such action **is having** the desired effect.]/

* subsequently：〔副詞〕その後，その次に　　　　* check：〔他動詞〕〜を確認する。
* that such action is having the desired effect：接続詞 that は「〜ということを」という意味で，that 以下は動詞⑩(shall) check の目的語である。
* is having：〔現在進行形〕現在，適切な効果がある
* desired：〔形容詞〕適切な，願っていた（effect を修飾している）
* effect：〔名詞〕効果，結果
* 接続詞⑦ and は何と何をつなぐか？→動詞⑥ shall also take と動詞⑩ (shall) check をつなぐ。

⑧⑨⑩部分訳 ＿＿＿＿＿＿＿＿＿＿＿＿＿＿＿＿＿＿＿＿＿＿＿＿＿＿＿＿＿＿＿＿＿＿＿＿＿

全文和訳

3) In clear weather, /whenever possible,/ the officer in charge of the navigational watch **shall carry out**
①　　　　　　　②　　　　　　　　　　　　　　③　　　　　　　　　　　　　　　⑤
radar practice./
④

* whenever (it is) possible：可能なときはいつでも　　　* carry out：〔熟語〕～を実行する
* radar practice：radar は電波が反射する時間・方向を測定して目標物の存在・位置を認知する装置
　practice はここでは「実行」の意味だが，「レーダーによる watch（当直）実施」という意味。

①②部分訳 _____

全文和訳
_____

Step 3 網掛けは主語，**ゴシック体**は動詞（述語動詞）

(1) 音声を聞きながら，音読する。音読が終わったら，Step Check Box □をぬりなさい。
(2) 左から順に，スラッシュ毎の意味を考える。

　The officer in charge of the navigational watch **shall take** frequent and accurate compass bearings/ of approaching ships/ as a means of early detection of risk of collision/ and **bear in mind**/ that such risk **may** sometimes **exist**/ even when an appreciable bearing change **is** evident,/ particularly when **approaching** a very large ship or a tow/ or when **approaching** a ship/ at close range./
　The officer in charge of the navigational watch **shall** also **take** early and positive action/ in compliance with the applicable International Regulations for Preventing Collisions at Sea, 1972/ and subsequently **check**/ that such action **is having** the desired effect./
　In clear weather,/ whenever possible,/ the officer in charge of the navigational watch **shall carry out** radar practice./

【文法のポイント】
後置修飾のパタンを理解することがポイントである。ここでは，ポイントを5つに分ける。海技士試験に出てきた英文を例に下記に説明する。スラッシュ毎に英文の内容を理解し，修飾の仕方のポイント5つを参考に，日本語らしい語順に訳すスキルが必要である。
1. ［接続詞＋主語S＋動詞V…］は，動詞Vにかかる。
2. （前置詞＋名詞（もの・コト））は，直前（直後）の名詞や動詞にかかる。
3. ［関係代名詞＋主語S＋動詞V…］は，直前の名詞にかかる。
4. 名詞＋過去分詞（Ved）の語順の場合，過去分詞（Ved）は，直前の名詞にかかる。
5. 名詞＋現在分詞（Ving）の語順の場合，現在分詞（Ving）は，直前の名詞にかかる。

具体例
◆［接続詞＋主語S＋動詞V…］は，［主語S＋動詞V…］のなかの動詞Vにかかる

例1　In clear weather, [whenever (it is) possible], the officer in charge of the navigational watch **shall carry out** radar practice.

　　* whenever（～するときはいつでも）は複合関係副詞だが，接続詞と同じ働きをする。
　　　［whenever (it is) possible］は動詞 shall carry out にかかる（を修飾する）ことがわかると，訳す順もわかる。

例2　such risk **may** sometimes **exist** [even when an appreciable bearing change **is** evident.]

　　* [even when an appreciable bearing change **is** evident] は，直前の動詞 may sometimes exist にかかる。
　　　even when は「～のときでさえ」と訳す。

◆（前置詞＋名詞（もの・コト））は，直前（直後）の名詞や動詞にかかる

例3　The officer (in charge of the navigational watch) **shall take** frequent and accurate compass bearings

　　　＊(in charge of the navigational watch) が，直前の名詞 the officer を修飾することがわかると，訳順がわかる。

例4　frequent and accurate compass bearings (of approaching ships)

　　　＊(of approaching ships) が，直前の名詞 bearings を修飾している。
　　　（前置詞＋名詞）は，最小の意味のまとまりを表す。働きは2つあり，副詞のように動詞を修飾するか，形容詞のように名詞を修飾するか，どちらかである。（前置詞＋名詞）の意味を考えて，直前にある名詞あるいは動詞の意味がわかれば，名詞にかかるか動詞にかかるか，判断できる。

例5　**shall take** frequent and accurate compass bearings of approaching ships (as a means of early detection of risk of collision)

　　　＊(as a means of early detection of risk of collision) は，前の動詞 shall take にかかる。

例6　**shall** also **take** early and positive action (in compliance with the applicable International Regulations for Preventing Collisions at Sea, 1972)

　　　＊(in compliance with…) が，前の動詞 shall also take にかかる。

【専門英語単語リスト】書いておぼえよう

発音したり，スペル練習したり，意味を調べたりしたらチェック欄の○をぬりなさい。
辞書を引いて単語の意味が多くある場合，英語の文脈から判断し，適切な意味を選びなさい。

英文	単語	品詞	意味を調べて書きなさい	発音しながらスペル練習しなさい	チェック
1)	frequent	形容詞			○○○
1)	accurate	形容詞			○○○
1)	bearing(s)	名詞			○○○
1)	approach	名詞			○○○
1)	risk	名詞			○○○
1)	collision / collide	名／動			○○○
1)	bear in mind	熟語			○○○
1)	exist	自動詞			○○○
2)	take action	熟語			○○○
2)	in compliance with	熟語			○○○
2)	applicable	形容詞			○○○
2)	regulation	名詞			○○○
2)	prevent	他動詞			○○○
2)	desired	形容詞			○○○
3)	possible	形容詞			○○○

Unit 3 　航行の安全　自動操舵の使用　　Step Check Box □ □ □

Step 1　下記の英文を読み，次の3つの活動をしなさい。終了後 Step Check Box □ をぬりなさい。

(1) 動詞（述語動詞）を○で囲む。動詞の意味を考えて，動詞のあとに続く内容を推測する。
(2) 主語（主部）に下線を引く。
(3) スラッシュを入れる。スラッシュを入れる場所は主に下記の3つ。
- コンマ(,)やピリオド(.)でスラッシュ(/)を入れる。音読するときはスラッシュで息つぎする。
- 接続詞の前にスラッシュを入れる。［接続詞＋主語＋動詞］は文としてまとまった意味を表す。
  ここでいう接続詞とは when, if, until, before, after など（and, but, or, so を含まない）。
- 原則として前置詞の前にスラッシュを入れる。（前置詞＋名詞）は，最小の意味のまとまりを表す。

1) In areas of high traffic density, in conditions of restricted visibility and in all other hazardous navigational situations where heading and/or track control systems are in use, it shall be possible to establish manual control of the ship's steering immediately.

2) In circumstances as above, the officer in charge of the navigational watch shall have available without delay the services of a qualified helmsperson who shall be ready at all times to take over steering control.

3) The change-over from automatic to manual steering and vice versa shall be made by or under the supervision of a responsible officer.

4) The manual steering shall be tested after prolonged use of heading and/or track control systems, and before entering areas where navigation demands special caution.

出典：2級海技士（航海）平成22年2月定期試験問題
SOLAS 条約 Chapter 5, Regulation 24 からの抜粋

Step 2　網掛け は主語，**ゴシック体**は動詞（述語動詞），＊語注・文法的留意事項，①②③は訳順番号

(1) 単語や熟語の説明を読みながら，部分訳を書きなさい。わからない単語の意味は調べなさい。
(2) 部分訳が終わったら，①②③の訳順番号を参考に，全文和訳を書きなさい。
　　全文和訳が終わったら，Step Check Box □ をぬりなさい。

1) In areas of high traffic density,／ in conditions of restricted visibility／
　　　　　① 　　　　　　　　　　　　　　　　　②

* in the area of high traffic density：輻輳海域で
* traffic：〔名詞〕通行，交通量
* in conditions of restricted visibility：視界制限状況で
* visibility：〔名詞〕視程，視界
* density：〔名詞〕密集，（人口など）込みぐあい
* restrict：〔他動詞〕
* visible：〔形容詞〕目に見えて

and in all other hazardous navigational situations／ [where heading and/or track control systems **are** in use,]／
　　　　　　　⑤　　　　　　　　　　　　　　　　　　　　　　　③　　　　　　　　　　　　　④

* hazardous：〔形容詞〕危険な = dangerous
* hazard：〔名詞〕（偶然性の強い）危険 = danger
* navigational：〔形容詞〕航海の
* navigation：〔名詞〕航海，航行
* situation：〔名詞〕状況，事態 = conditions
* A and/or B：A そして B，あるいは A か B か
* heading (control system)：オートパイロット = automatic pilot
* track control system：トラックコントロールシステム
* be in use：〔熟語〕使用される
* where：〔関係副詞〕［where 〜 in use］が，直前の名詞 situations にかかる。「どこに」という意味はない。

①②部分訳 _____

③④⑤部分訳　オートパイロットとトラックコントロールシステムの両方，あるいはそのどちらか一方が使用されているその他の危険航行状況においては

16

　　　　　　　　⑦　　　　　　　　　　　　　　　　⑥
　　it **shall be** possible/ to establish manual control of the ship's steering immediately./

* it：仮主語なので「それは」と訳さない。to 以下が真の主語なので、「〜することは」と訳す。
* 主語 S と動詞 V がわかると、トピックの内容を絞りこむことが容易になる。ここでは、「船舶操舵の手動制御への切り替え」が、トピックである。
* shall be possible：当然ありえることだ
* establish：〔他動詞〕〜を設定する
* ship's steering：船舶の操舵
* possible：〔形容詞〕可能な、実行できる ↔ 反対語 impossible
* manual control：〔名詞〕手動制御
* immediately：〔副詞〕ただちに = soon

⑥部分訳 _____ することは

全文和訳 _____

　　　　　　　①　　　　　　　③　　　　　　　②　　　　　　　⑧
2) In circumstances as above,/ the officer in charge of the navigational watch **shall have**

* circumstance：〔名詞〕事情
* as above：上述のような
* in charge of：責任のある
* shall have 目的語 available：have は使役動詞で「〜させる」の意味。
　ここでは「(④⑤⑥＝目的語)を利用すべきである(＝するようにすべきである)」となる。

②③⑧部分訳 _____

　　⑧　　　　　⑦　　　　　　　⑥　　　　　　　　　　　⑤　　　　　　　　　　　④
available/ without delay/ the services of a qualified helmsperson/ [who **shall be ready** at all times **to** take over
④
steering control.]/

* available：〔形容詞〕利用できて
* service：〔名詞〕業務・任務
* who：[関係代名詞 who 以下] は a qualified helmsperson を修飾する働きをしている。
* at all times：いつでも
* take over：引き継ぐ
* without delay：遅れることなく
* a qualified helmsperson：資格をもった操舵手
* be ready to V：すぐに〜できる、〜する用意がある
* steering control：操舵制御

④⑤⑥部分訳　いつでもすぐに _____ できる資格をもった _____

全文和訳 _____

　　　　　②　　　　　　　　　　　　①　　　　　　　　　　　　③　　　　　　⑥
3) The change-over /from automatic (steering) to manual steering/ and vice versa **shall be made**/

* change-over：切り替え、転換
* automatic steering：自動操舵
* vice versa：またはその逆、「バイスバーサ」と読む。
* shall be made：〔受動態〕なされるべきだ
* from A to B：A から B へ
* manual steering：手動操舵

①②③部分訳 _____

　　　　⑤　　　　　　　　　　　　④
by or under the supervision/ of a responsible officer./

* by or under：〜によってあるいはその下で
* supervision：〔名詞〕監督
* responsible：〔形容詞〕責任のある

④⑤部分訳 _____

全文和訳 _____

4) The manual steering **shall be tested**/ after prolonged use of heading and/or track control systems,/
⑤ ⑥ ① ②

* shall be tested：〔受動態〕検査されるべきである * steering：〔名詞〕操舵，操縦
* after prolonged use 〜 control systems：直前の動詞 shall be tested にかかる。
* prolonged use：長期の使用 * prolonged：〔形容詞〕= long

①②部分訳 _____ の後で

and before entering areas/ [where **navigation demands** special caution.]/
④ ③

* before：〔前置詞〕〜する前に * area：〔名詞〕海域
* enter：〔他動詞〕〜に入る * (before) entering areas：〜の海域に入る前に
* where：〔関係副詞〕[where navigation demands special caution] は，areas にかかる。
* demand：〔他動詞〕〜を要求する * caution：〔名詞〕注意

③④部分訳 _____ 前に

全文和訳 _____

Step 3   網掛け は主語，**ゴシック体**は動詞（述語動詞）

(1) 音声を聞きながら，音読する。音読が終わったら，Step Check Box □をぬりなさい。
(2) 左から順に，スラッシュ毎の意味を考える。

In areas of high traffic density,/ in conditions of restricted visibility/ and in all other hazardous navigational situations/ where heading and/or track control systems **are** in use,/ it **shall be** possible/ to establish manual control of the ship's steering immediately./

In circumstances as above,/ the officer in charge of the navigational watch **shall have** available/ without delay/ the services of a qualified helmsperson/ who **shall be ready** at all times **to** take over steering control./

The change-over/ from automatic to manual steering/ and vice versa **shall be made**/ by or under the supervision/ of a responsible officer./

The manual steering **shall be tested**/ after prolonged use of heading and/or track control systems,/ and before entering areas/ where navigation **demands** special caution./

【文法のポイント】
◆ （前置詞＋名詞（もの・コト））は，直前（直後）の名詞や動詞にかかる

例1 (In areas of high traffic density), (in conditions of restricted visibility) and (in all other hazardous navigational situations)

＊前置詞は，他の単語と一緒に熟語として使われる場合がある。ここでは，前置詞 in が3回続けて使われている。in は「容器のなかにある」というイメージなので，「〜という場所（のなか）で」「〜という状況（のなか）で」「〜という事態（のなか）で」という意味だと理解できる。ここでは（前置詞＋名詞）は，直後の動詞 shall be にかかる。

例2 The manual steering **shall be tested** (after prolonged use of heading and/or track control systems), and (before entering areas) [where navigation **demands** special caution.]

＊after と before の後には prolonged use（名詞）と entering（動名詞）が続いているので，after と before は前置詞として機能している。
（after prolonged use of heading and/or track control systems）と（before entering areas）は，「〜のあとに」と「〜のまえに」というように対比して書かれているので，わかりやすい。
＊長文の内容理解には，このように前置詞や接続詞がどのように使われているのかに着目すると助けになる。

◆仮主語 it と真の主語が to 不定詞以下の場合

例3　it **shall be** possible (to establish manual control of the ship's steering immediately.)

　　＊it は仮主語，真の主語は to 以下である。to 以下には S や V は不要である。to + V（動詞の原形）は，「～することは」と訳す。

例4　it **may be** necessary (to transfer all or part of her cargo to another vessel.)

　　＊it は仮主語，真の主語は to 以下である。to 以下には S や V は不要である。to + V（動詞の原形）は，「～することは」と訳す。

◆仮主語 it と真の主語が that 以下の場合

例5　It **is** important [that all means for indicating the position of ships in distress or survival craft **should be** properly **used**.]

　　＊it は仮主語，that 以下が真の主語なので，that は接続詞（～ということ）である。that 以下には主語 S と動詞 V がある。主語 S は all means for indicating the position of ships in distress or survival craft であり，動詞は should be properly used である。「～が…するということは」と訳す。
　　＊that の直前に名詞がくる関係代名詞 that と違って，接続詞 that の直前には形容詞や動詞がくる。

【専門英語単語リスト】書いておぼえよう

発音したり，スペル練習したり，意味を調べたりしたらチェック欄の○をぬりなさい。
辞書を引いて単語の意味が多くある場合，英語の文脈から判断し，適切な意味を選びなさい。

英文	単語	品詞	意味を調べて書きなさい	発音しながらスペル練習しなさい	チェック
1)	area	名詞			○○○
1)	high traffic density	熟語			○○○
1)	restricted / restrict	形 / 動			○○○
1)	visibility / visible	名 / 形			○○○
1)	heading (control system)	名詞			○○○
1)	track control system	熟語			○○○
1)	steering	名詞			○○○
2)	service	名詞			○○○
2)	at all times	熟語			○○○
2)	take over	熟語			○○○
3)	manual steering	熟語			○○○
4)	prolonged	形容詞			○○○
4)	navigation / navigate	名 / 動			○○○
4)	demand	他動詞			○○○
4)	caution	名詞			○○○

Unit 4　海上労働　年少船員

Step Check Box □ □ □

Step 1　下記の英文を読み，次の3つの活動をしなさい。終了後 Step Check Box □をぬりなさい。

(1) 動詞（述語動詞）を○で囲む。動詞の意味を考えて，動詞のあとに続く内容を推測する。
(2) 主語（主部）に下線を引く。
(3) スラッシュを入れる。スラッシュを入れる場所は主に下記の3つ。
- コンマ(,)やピリオド(.)でスラッシュ(/)を入れる。音読するときは，スラッシュで息つぎする。
- 接続詞の前にスラッシュを入れる。［接続詞＋主語＋動詞］は文としてまとまった意味を表す。
  ここでいう接続詞とは when, if, until, before, after など（and, but, or, so を含まない）。
- 原則として前置詞の前にスラッシュを入れる。（前置詞＋名詞）は，最小の意味のまとまりを表す。

1) Standard A1.1-Minimum age
2) The employment, engagement or work on board a ship of any person under the age of 16 shall be prohibited.
3) Night work of seafarers under the age of 18 shall be prohibited.
4) For the purposes of this Standard, "night" shall be defined in accordance with national law and practice.
5) It shall cover a period of at least nine hours starting no later than midnight and ending no earlier than 5 a.m.
6) An exception to strict compliance with the night work restriction may be made by the competent authority when:
6-1) The effective training of the seafarers concerned, in accordance with established programmes and schedules, would be impaired; or
6-2) The specific nature of the duty or a recognized training programme requires that the seafarers covered by the exception perform duties at night and the authority determines, after consultation with the shipowners' and seafarers' organizations concerned, that the work will not be detrimental to their health or well-being.

出典：2級海技士（航海）平成 25 年 10 月定期試験問題
Maritime Labour Convention, 2006（2006 年の海上の労働に関する条約）A1.1 からの抜粋

Step 2　網掛けは主語，ゴシック体は動詞（述語動詞），＊語注・文法的留意事項，①②③は訳順番号

(1) 単語や熟語の説明を読みながら，部分訳を書きなさい。わからない単語の意味は調べなさい。
(2) 部分訳が終わったら，①②③の訳順番号を参考に，全文和訳を書きなさい。
　全文和訳が終わったら，Step Check Box □をぬりなさい。

1) Standard A1.1-Minimum age

　＊ standard：標準，おきて，基準。ここでは「規律，規則」という意味。
　＊ A1.1：A は Article「条項」を意味する。
　＊「最低年齢」がトピックなので，雇用条件などの知識が問われる。

タイトル和訳

2) The employment, engagement or work／ on board a ship／ of any person under the age of 16 **shall be prohibited**.／
　　　　　　　　　③　　　　　　　　　　　　　②　　　　　　　　　①　　　　　　　　　　　　　　　④

　＊ employment：〔名詞〕雇用　　　　　　　　　　＊ employ：〔他動詞〕
　＊ engagement：〔名詞〕一定期間の契約　　　　＊ engage：〔他動詞〕
　＊ on board：〔前置詞〕（船などに）乗って　　　＊ on board a ship：船舶において
　＊ the employment, engagement or work on board a ship：①の船舶においての雇用，契約雇用あるいは労働は
　＊ under：〔前置詞〕より下の（ここでは「16 歳より下の」「16 歳未満の」（16 歳は含まない）を意味する）
　＊ shall be prohibited：〔受動態〕禁止されるべきである（海技士試験では「助動詞＋受動態」の形の動詞が多い）

全文和訳

3) Night work of seafarers under the age of 18 **shall be prohibited**./
   ③ ② ①

   * seafarer：〔名詞〕船員　　　* night work：夜勤　　　* prohibit：〔他動詞〕～を禁止する
   * 条文なので，助動詞 shall be Ved〔受動態〕が頻繁に使用されている。「～されるべきだ」という意味。

   全文和訳
   _____

4) For the purposes of this Standard,/ "night" **shall be defined**/ in accordance with national law and practice./
   ② ① ④ ③

   * for the purpose of ～：〔熟語〕～の目的のために，～のために　　* purpose：〔名詞〕目的
   * define：〔他動詞〕定義する　　　　　　　　　　　　　　　　* shall be defined：〔受動態〕定義されるべきである
   * in accordance with：〔熟語〕～に従い，～のとおりに
   * national law and practice：国内法と慣例

   全文和訳
   _____

5) It **shall cover** a period of at least nine hours/ starting no later than midnight/ and ending no earlier than 5 a.m./
   ① ⑤ ④ ② ③

   * It：前の英文の主語 night を指す。　　　　　　　　* cover：〔他動詞〕を含む，扱う，包む（覆う）
   * period：〔名詞〕期間　　　　　　　　　　　　　　* a period of ～：～の期間
   * at least：〔熟語〕少なくとも　　　　　　　　　　　* little-less-least：少ない－より少ない－最も少ない
   * no later than：〔熟語〕遅くても～以内に，～までに　* late-later-latest：時刻が遅い－もっと遅い－最も遅い
   * starting no later than midnight：遅くとも真夜中の12時までに始まる。starting 以下は period にかかる。
   * ending no earlier than 5 a.m.：早くても朝5時以降に終わる。ending 以下は period にかかる。
   * 接続詞 and は何と何をつなぐか？→ ②と③をつなぐ。

   全文和訳
   _____

6) An exception to strict compliance with the night work restriction **may be made**/ by the competent authority/
   ③ ② ① ⑥ ⑤
   when:
   ④

   * exception：〔名詞〕例外，特例　　　　　　　　　　* except：〔前置詞〕～を除いては，以外は
   * to strict compliance with：①への厳格な遵守に対する
   * restriction：〔名詞〕制限　　　　　　　　　　　　* restrict：〔他動詞〕
   * competent：〔形容詞〕管轄権を有する，有能な　　　* authority：〔名詞〕機関
   * when：この when は，「以下のときに，①への厳格な遵守に対する例外が…によって出される」と訳す。
     「以下」とは，6-1) 6-2) の英文を指す。competent authority による「例外の条件」が述べられている。
   * may be made：may〔助動詞〕＋受動態。may は「～しなければいけない」「～するものとする」の意味。

   ①②③部分訳 _____

   全文和訳
   _____

6-1) The effective training of the seafarers concerned,/ in accordance with established programmes and schedules,/
     ④ ③ ② ①
     would be impaired;/ or
     ⑤

   * effective：〔形容詞〕実践的な，効果的な，有効な　　* effect：〔名詞〕
   * concerned：〔形容詞〕当該の，関連している。seafarers にかかる。
   * in accordance with：〔熟語〕～に従って，～のとおりに　* accordance：〔名詞〕調和，許可
   * established：〔形容詞〕制定された，定評のある
   * programme：プログラム。イギリス英語のスペル。アメリカ英語のスペルは program である。
   * be impaired：十分に機能（役目）を果たさない，正常に働かない
   * 接続詞 or は何と何をつないでいるか？→「6-1) あるいは 6-2)」と訳し，これら2文をつなげる。

   全文和訳
   _____のとき，あるいは

6-2) The specific nature of the duty/ or a recognized training programme **requires**/

* specific nature：特有の性質，特質　　　　　　　　* duty：〔名詞〕任務
* 接続詞 or は何と何を結んでいるか？→「①あるいは②」と訳し，①と②をつなげている。
* recognized training programme：公認の訓練プログラム。スペルに注意：programme（英語），program（米語）。
* 海技士試験では，イギリス英語のスペルを使用することが多い。例：centre（英語），center（米語）。
* require：〔他動詞〕〔that 以下③④⑤⑥のこと〕を義務付ける

①②部分訳 _____

[that the seafarers covered by the exception (should) **perform** duties at night]/

* that：〔接続詞〕～ということを。that 以下には，主語Sと動詞V（should perform）がある。
* covered by the exception：例外として扱われている。the seafarers を修飾している。
* (should) perform：〔他動詞〕果たす。should が省略されている。* performance：〔名詞〕
* duties at night：夜勤

and the authority **determines**,/

* authority：〔名詞〕機関　　　　　　　　　　　* determine：〔他動詞〕～を決定する＝firmly decide
* determines の目的語は，[that 以下⑬⑭⑮⑯] である。
* 接続詞 and は何と何をつないでいるか？→①②⑦と⑨⑰をつないでいる。

⑨⑰部分訳　そして _____

after consultation with the shipowners' (organizations) and seafarers' organizations concerned,/

* after consultation with ～：〔熟語〕AとBに相談した後で
* concerned：〔形容詞〕関係している，直前の名詞 organizations を修飾する。
* shipowners' (organizations) and seafarers' organizations：船舶所有者および船員の組織

⑩⑪⑫部分訳 _____

[that the work will not be detrimental/ to their health or well-being.]/

* that 主語S＋動詞V：〔接続詞〕～ということを　　* detrimental：〔形容詞〕（健康などに）有害な
* be detrimental to ～：～にとって有害である　　　* well-being：〔名詞〕福祉

⑬⑭⑮⑯部分訳 _____

全文和訳

Step 3　網掛け は主語，**ゴシック体**は動詞（述語動詞）

(1) 音声を聞きながら，音読する。音読が終わったら，Step Check Box □をぬりなさい。
(2) 左から順に，スラッシュ毎の意味を考える。

Standard A1.1-Minimum age

1　The employment, engagement or work/ on board a ship/ of any person under the age of 16 **shall be prohibited**./

2　Night work of seafarers under the age of 18 **shall be prohibited**./ For the purposes of this Standard,/ "night" **shall be defined**/ in accordance with national law and practice./ It **shall cover** a period of at least nine hours/ starting no later than midnight/ and ending no earlier than 5 a.m./

3 An exception to strict compliance with the night work restriction **may be made**/ by the competent authority/ when:
   (a) The effective training of the seafarers concerned,/ in accordance with established programmes and schedules,/ **would be impaired**;/ or
   (b) The specific nature of the duty/ or a recognized training programme **requires**/ that the seafarers covered by the exception **perform** duties at night/ and the authority **determines**,/ after consultation with the shipowners' and seafarers' organizations concerned,/ that the work **will not be** detrimental/ to their health or well-being./

【専門英語単語リスト】書いておぼえよう

発音したり，スペル練習したり，意味を調べたりしたらチェック欄の○をぬりなさい。
辞書を引いて単語の意味が多くある場合，英語の文脈から判断し，適切な意味を選びなさい。

英文	単語	品詞	意味を調べて書きなさい	発音しながらスペル練習しなさい	チェック
1)	standard	名詞			○○○
2)	prohibit	他動詞			○○○
3)	seafarer	名詞			○○○
4)	purpose	名詞			○○○
4)	define	他動詞			○○○
4)	in accordance with	熟語			○○○
4)	practice	名詞			○○○
5)	cover	他動詞			○○○
6)	compliance	名詞			○○○
6)	restriction	名詞			○○○
6)	authority	名詞			○○○
6-2)	duty	名詞			○○○
6-2)	recognize	他動詞			○○○
6-2)	require	他動詞			○○○
6-2)	perform	他動詞			○○○

理解を深めるキーワード「年少船員」

船員の労働に関する国際条約に，MLC条約（海上の労働に関する国際条約；Maritime Labour Convention, 2006）がある。この国際条約はILO（国際労働機関；International Labour Organization）において採択された条約である。このUnitで取り扱った船員の最低年齢や，年少船員に禁止される作業内容は，この条約のなかで規定されている。この他に，船員の雇用や労働条件，居住や厚生施設，食事，保険，医療，福利厚生，社会保障などに対する要件が規定されている。これらを規定することで，労働者としての船員を危険で過酷な労働から保護し，労働環境や労働条件の改善を図ることで人間的な保護を与えることを目的としている。労働に関しては，国内法では船員法で取り扱われている。

Unit 5　船員の訓練 1

Step Check Box □ □ □

Step 1　下記の英文を読み，次の3つの活動をしなさい。終了後 Step Check Box □をぬりなさい。

(1) 動詞（述語動詞）を○で囲む。動詞の意味を考えて，動詞のあとに続く内容を推測する。
(2) 主語（主部）に下線を引く。
(3) スラッシュを入れる。スラッシュを入れる場所は主に下記の3つ。
- コンマ (,) やピリオド (.) でスラッシュ (/) を入れる。音読するときは，スラッシュで息つぎする。
- 接続詞の前にスラッシュを入れる。[接続詞＋主語＋動詞]は文としてまとまった意味を表す。
  ここでいう接続詞とは when, if, until, before, after など（and, but, or, so を含まない）。
- 原則として前置詞の前にスラッシュを入れる。（前置詞＋名詞）は，最小の意味のまとまりを表す。

1) Training of officers and ratings serving on oil and chemical tankers
2) Officers and ratings having specific duties in connection with the cargo and cargo equipment should undergo training divided into two parts.
3) A general part concerning the principles involved and the other part on the application of those principles to ship operation.
4) Any of this training may be given by properly qualified personnel.
5) All other persons employed on these ships should undergo training on board ships and, where appropriate, ashore, which should be given by qualified personnel experienced in the characteristics and handling of oil and chemical cargoes and safety procedures.

出典：2級海技士（航海）平成26年4月定期試験問題

Step 2　網掛けは主語，**ゴシック体**は動詞（述語動詞），＊語注・文法的留意事項，①②③は訳順番号

(1) 単語や熟語の説明を読みながら，部分訳を書きなさい。わからない単語の意味は調べなさい。
(2) 部分訳が終わったら，①②③の訳順番号を参考に，全文和訳を書きなさい。
　全文和訳が終わったら，Step Check Box □をぬりなさい。

1) Training of officers and ratings/ serving on oil (tankers) and chemical tankers
　　　　　③　　　　　　　②　　　　　　　　　　　　　　　①

- ＊ training：〔名詞〕訓練　　　＊ officer：〔名詞〕士官　　　＊ rating：〔名詞〕部員
- ＊ serving：〔現在分詞〕serving 以下は officers and rating にかかる。「～の仕事をしている」の意味。
- ＊ chemical：〔形容詞〕化学の，〔名詞〕化学薬品，化学製品
- ＊ oil tanker：オイルタンカー，原油タンカー
- ＊ chemical tanker：ケミカルタンカー
- ＊ Training が「主題」（最も言いたい内容）である。英語では，主題を最初に述べ，次に詳細を述べるという順番で書かれることが多い。

タイトル和訳＿＿＿＿＿＿＿＿＿＿＿＿＿＿＿＿＿＿＿＿＿＿＿＿＿＿＿＿＿＿＿＿

2) Officers and ratings/ having specific duties/ in connection with the cargo and cargo equipment
　　　④　　　　　　　　③　　　　　　　　　②　　　　　　　　　①
   should undergo training /divided into two parts./
　　　⑥　　　　　　　　⑤

- ＊ having：〔現在分詞〕having 以下は officers and ratings にかかる。　　＊ specific duties：特定の任務
- ＊ in connection with：〔熟語〕～に関連している
- ＊ the cargo and cargo equipment：船荷や積荷装置。ここでは oil and chemical（原油や化学薬品関連）である。
- ＊ undergo：〔他動詞〕経験する
- ＊ divided into two parts：〔過去分詞〕divided 以下は直前の名詞 training にかかる。
  ⑤⑥の大意は「2つの部分からなる訓練を経験すべきだ」となる。
  two parts と書いてあるので，次にそれを説明する文が書いてあると予測できる。

①②③部分訳＿＿＿＿＿＿＿＿＿＿＿＿＿＿＿＿＿＿＿＿＿＿＿＿＿＿＿＿＿＿＿＿

全文和訳＿＿＿＿＿＿＿＿＿＿＿＿＿＿＿＿＿＿＿＿＿＿＿＿＿＿＿＿＿＿＿＿

3) A general part/ concerning the principles (is) involved/ and the other part (is)/ on the application of those principles/ to ship operation./

- ＊前文2）で two parts と書いてあるので，2つの訓練 a general part と the other part が述べられている。
- ＊ general：〔形容詞〕一般的な，全体的な ＊ concerning：〔前置詞〕〜に関して＝about
- ＊ principle：〔名詞〕根本方針，原則 ＊ (is) involved：〔受動態〕含まれる，〜が必要とされる
- ＊①②③の大意は「根本原則に関する全体的な訓練が必要である」となる。
- ＊ the other part is on the application：もう一つ別の訓練は，〜することについてである
- ＊ on the application of those principles to 〜：those principles を to 以下に応用すること。
- ＊ to ship operation：船舶運航に
- ＊ those principles：それらの原則 ＊ those：〔代名詞〕that の複数形
- ＊④⑤⑥⑦の大意は「もう一つ別の訓練は，船舶運航にそれらの原則を応用することについてである」となる。

全文和訳

4) Any of this training may be given/ by properly qualified personnel./

- ＊ any of this training：この訓練のどれも ＊ may：〔助動詞〕〜しなければならない
- ＊ properly qualified personnel：適切に資格を有する人材
- ＊ properly：〔副詞〕正しく ＊ proper：〔形容詞〕
- ＊ qualified：〔形容詞〕資格のある ＊ qualify：〔他動詞〕資格を与える

全文和訳

5) All other persons/ employed on these ships should undergo training/ on board ships/

- ＊ all other persons employed on these ships：これらの船に雇われている他のすべての人は
- ＊ employed：〔過去分詞〕直前の名詞 all other persons にかかる。
  ここでは，should undergo が動詞なので，employed は動詞の過去形ではなくて，過去分詞形だと判断できる。
- ＊ undergo：〔他動詞〕〜を経験する
- ＊ other persons：「他の」というのは，上官と部員 (officers and ratings) 以外の他の乗組員を意味する。
- ＊ training on board ships：乗船訓練（陸上での訓練は下記英文⑤〜⑮で述べられている）

①②③④部分訳 _____

and, (at the place) [where (it is) appropriate,/ ashore],/(and it is training) which should be given/

- ＊ at the place where it is appropriate, ashore：先行詞 (at the place) が省略されている。
  where は関係副詞。where 以下が at the place にかかる。「where 以下のところでは」と訳す。
- ＊ ashore：〔副詞〕陸上で。on board ships and ashore（乗船訓練そして陸上での訓練）と意味がつながる。
- ＊⑤⑥⑦の大意は「そして，陸上の適切な場所での訓練」となる。
- ＊ (and it is training) which should be given：そしてそれは⑬以下によって与えられるべき訓練である

by qualified personnel experienced in the characteristics and handling of oil and chemical cargoes/ and safety procedures./

- ＊ by qualified personnel experienced in 〜：〜の経験がある有資格者による
- ＊ experienced in 〜：〔形容詞〕〜の経験がある
- ＊ characteristics：〔名詞〕特徴 ＊⑧⑨⑩⑪大意：原油と化学薬品の積荷の特徴と移動の方法そして安全手順
- ＊ handling：〔名詞〕移動の方法，操縦，取扱い ＊ oil and chemical cargoes：原油と化学薬品の積荷
- ＊ safety procedure：安全手順（手続き）

⑧⑨⑩⑪⑫⑬部分訳 _____

全文和訳

Step 3 網掛けは主語，**ゴシック体**は動詞（述語動詞）

(1) 音声を聞きながら，音読する。音読が終わったら，Step Check Box □をぬりなさい。
(2) 左から順に，スラッシュ毎の意味を考える。

Training of officers and ratings/ serving on oil and chemical tankers

1. Officers and ratings/ having specific duties/ in connection with the cargo and cargo equipment **should undergo** training/ divided into two parts./ A general part/ concerning the principles **involved**/ and the other part/ on the application of those principles/ to ship operation./ Any of this training **may be given**/ by properly qualified personnel./

2. All other persons/ employed on these ships **should undergo** training/ on board ships/ and, where appropriate,/ ashore,/ which **should be given**/ by qualified personnel experienced in the characteristics and handling of oil and chemical cargoes/ and safety procedures./

【文法のポイント】
スラッシュ毎に立ち止まって訳しながら，英文の内容を理解する。次に，和訳を書くときは，下記のポイントを参考に，日本語らしい語順に直す。

1. ［接続詞＋主語Ｓ＋動詞Ｖ…］は，動詞Ｖにかかる。
2. （前置詞＋名詞（もの・コト））は，直前（直後）の名詞や動詞にかかる。
3. ［関係代名詞＋主語Ｓ＋動詞Ｖ…］は，直前の名詞にかかる。
4. 名詞＋過去分詞（Ved）の語順の場合，過去分詞（Ved）は，直前の名詞にかかる。
5. 名詞＋現在分詞（Ving）の語順の場合，現在分詞（Ving）は，直前の名詞にかかる。

具体例
◆（前置詞＋名詞（もの・コト））は，直前（直後）の名詞や動詞にかかる

例1　A general part (concerning the principles)
　　＊（前置詞＋名詞）は，最小の意味のまとまりを表し，形容詞のように直前の名詞にかかる（を修飾する）

例2　Officers and ratings having specific duties (in connection with the cargo and cargo equipment)

【合格のためのポイント】
- 海技士試験は，辞書を持ち込むことができる。日ごろの英語の勉強でも，面倒がらずに，辞書を引く練習をする。
- 長文読解のポイントは，述語動詞を見つけることである。受動態や助動詞 shall, should が多く使用されるので，（助動詞＋動詞の原形）（be動詞＋Ved）（助動詞＋be動詞＋Ved）のパタンに注意すると，述語動詞を見つけやすい。

【専門英語単語リスト】書いておぼえよう

発音したり，スペル練習したり，意味を調べたりしたらチェック欄の○をぬりなさい。
辞書を引いて単語の意味が多くある場合，英語の文脈から判断し，適切な意味を選びなさい。

英文	単語	品詞	意味を調べて書きなさい	発音しながらスペル練習しなさい	チェック
1)	train	他動詞			○○○
1)	serve / service	動 / 名			○○○
1)	chemical	形 / 名			○○○
2)	specific	形容詞			○○○
2)	undergo	他動詞			○○○
2)	divide	他動詞			○○○
3)	principle	名詞			○○○
3)	application / apply	名 / 動			○○○
3)	operation / operate	名 / 動			○○○
4)	qualify / qualified	動 / 形			○○○
5)	employ	他動詞			○○○
5)	experienced	形容詞			○○○
5)	characteristic	名詞			○○○
5)	handle	他動詞			○○○
5)	procedure	名詞			○○○

Unit 6 船員の訓練2

Step Check Box □ □ □

Step 1 下記の英文を読み，次の3つの活動をしなさい。終了後 Step Check Box □ をぬりなさい。

(1) 動詞（述語動詞）を○で囲む。動詞の意味を考えて，動詞のあとに続く内容を推測する。
(2) 主語（主部）に下線を引く。
(3) スラッシュを入れる。スラッシュを入れる場所は主に下記の3つ。
- コンマ (,) やピリオド (.) でスラッシュ (/) を入れる。音読するときは，スラッシュで息つぎする。
- 接続詞の前にスラッシュを入れる。〔接続詞＋主語＋動詞〕は文としてまとまった意味を表す。
  ここでいう接続詞とは when, if, until, before, after など (and, but, or, so を含まない)。
- 原則として前置詞の前にスラッシュを入れる。（前置詞＋名詞）は，最小の意味のまとまりを表す。

1) Guidance regarding additional training for masters and chief mates of large ships and ships with unusual maneuvering characteristics

2) It is important that masters and chief mates should have had relevant experience and training before assuming the duties of master or chief mates of large ships or ships having unusual maneuvering and handling characteristics significantly different from those in which they have recently served.

3) Such characteristics will generally be found in ships which are of considerable deadweight or length of special design or of high speed.

4) Prior to their appointment to such a ship, masters and chief mates should:

5) be informed of the ship's handling characteristics by the company, particularly in relation to the knowledge, understanding and proficiency listed under ship maneuvering and handling in column 2 of table A-II/2—

6) Specification of the minimum standard of competence for masters and chief mates of ships of 500 gross tonnage or more; and

7) be made thoroughly familiar with the use of all navigational and maneuvering aids fitted in the ship concerned, including their capabilities and limitations.

出典：1級海技士（航海）平成24年2月定期試験問題
STCW条約　STCWコード　B-V/a からの抜粋

Step 2　網掛けは主語，ゴシック体は動詞（述語動詞），＊語注・文法的留意事項，①②③は訳順番号

(1) 単語や熟語の説明を読みながら，部分訳を書きなさい。わからない単語の意味は調べなさい。
(2) 部分訳が終わったら，①②③の訳順番号を参考に，全文和訳を書きなさい。
　　全文和訳が終わったら，Step Check Box □ をぬりなさい。

1) Guidance/ ⑤ regarding additional training/ ④ for masters and chief mates/ ③ of large ships and ships/ ② with unusual ① maneuvering characteristics/

* guidance：〔名詞〕指針，指導，案内
* additional：〔形容詞〕付加的な，追加の → add〔動詞〕
* for masters：〔名詞〕船長のための
* for：〔前置詞〕〜のための
* unusual：〔形容詞〕他とは違う，普通ではない ⇔ usual〔形容詞〕
* maneuvering characteristics：操縦性能
* maneuver：〔他動詞〕〜を操縦する（manoeuvre と書かれている場合はイギリス英語のスペルである）
* guidance（案内書）なので，regarding 以下①②③④を読み，2) 以下の英文内容を推測する。
* regarding：〔前置詞〕〜に関して＝concerning
* chief mate (officer)：一等航海士
* with：〔前置詞〕〜を持っている，〜で

タイトル和訳

2) It **is** important/ that masters and chief mates **should have had** relevant experience and training/
　　⑪　　　　　　⑨　　　　　　　　　⑩　　　　　　　⑧

* It は仮主語 that masters and chief mates … in which they have recently served 全文が真の主語である。It は訳さない。
  that 以下にも主語 S (masters and chief mates) と動詞 V (should have had) がある。

28

* should + have + Ved（過去分詞）：過去の事実についての現在の気持ちを伝える。
　　　　　　　　　　　　　ここでは「出港前に訓練をするべきである」と訳す。
* relevant：〔形容詞〕関連性のある　　　　　* experience and training：経験と訓練

⑧⑨⑩⑪部分訳 _____

　　　　　　　　　⑦　　　　　　　⑥　　　　　　　⑤　　　　　　　　　　　④　　　　　　　　　③
before assuming the duties/ of master or chief mate/ of large ships or ships/ having unusual maneuvering and
handling characteristics/ significantly different from those /[in which they **have** recently **served**.]}/
　　　　　　　　　　　　　　　　　　　②　　　　　　　　　　　　　　　　　①

* assume：〔他動詞〕(役目・任務など)を引き継ぐ　(before〔前置詞〕の後なので assuming〔動名詞〕の形にする)
* duties：〔名詞〕勤め，職務
* master or chief mate of large ships or ships：巨大船や他の船の船長あるいは一等航海士
* having：〔現在分詞〕持っている　(having 以下は large ships or ships を修飾している)
* maneuvering and handling：両方とも「操船する」という意味。　* handling characteristics：操船性能
* ③④の大意は「巨大船や，他と異なる操縦性能を持つ船」となる。
* significantly：〔副詞〕著しく，重要なことには
* significantly different from ～：～とは著しく異なる
* those：〔代名詞〕that の複数名詞で，ここでは前出の複数名詞 large ships or ships を指す。
* in which they have recently served：they は masters and chief mates を指す。in which 以下は直前の代名詞 those にかかる。
* ①②の大意は「彼らが最近まで任務を果たしていた船とは著しく異なっている」となる。
* ②は，③の characteristics にかかる。　　　　　*③は，④の large ships and ships にかかる。

③④⑤⑥⑦部分訳 _____

_____ を引き継ぐ前に

全文和訳 _____

　　　　　　　　　　　　①　　　　　　　　　　　　　⑥　　　　　⑤　　　　　　　　　　　　③
3) Such characteristics **will** generally **be found** in ships/ [which **are** of considerable deadweight or length
　　　　　②　　　　　　　　　　　　④
of special design/ or (are) of high speed.]/

* such characteristics：そのような特徴　(前文2)の③から，どのような特徴かという説明が続く)
* generally：〔副詞〕一般的に → general〔形容詞〕　　*形容詞に ly が付くと副詞になることが多い。
* considerable：〔形容詞〕かなりの　　　　　　　　　* dead weight：載荷重量，積載重量
* are of considerable deadweight or (be of) length：かなりの載荷重量か，あるいは長さがある
* of special design：特別設計をした　　　　　　　* of + length = long
* or of high speed：あるいは高速力のある　　　　　* (of + 名詞)：形容詞の働きをする。
* [which 以下の英文] は，直前の名詞 ships にかかる。which 以下には船の特徴が書かれている。

②③④⑤部分訳 _____

全文和訳 _____

　　　　　　　　　　　　①　　　　　　　　　　　　　　　　　　②
4) Prior to their appointment to such a ship,/ masters and chief mates **should**:

* prior to：〔前置詞〕～より先だって，より前に = before
* appointment：〔名詞〕任用，任命
* prior to their appointment to such a ship：そのような船に任命される前に
* masters and chief mates should (be informed of ～ and be made ～)：主語②に対する動詞が5)と7)へと続く。
　should の後に「：」が付いていて，これを「コロン」という。具体例が後に続くということを示す。

全文和訳 _____

5) (should) **be informed of** the ship's handling characteristics/ by the company,/

* be informed of ～：〔動詞〕（～の事実）を知らされる
* handling：〔名詞〕操船，操縦，移動（輸送）方法，取り扱い
* characteristic：〔名詞〕特徴，特性
* by the company：会社から
* handling characteristics：操船性能

⑦⑧部分訳 _____

particularly in relation to the knowledge, understanding and proficiency/ listed under ship maneuvering and handling/ in column 2 of table A-Ⅱ/2―

* particularly：〔副詞〕in relation to にかかり，「とくに～に関する」と訳す。⑤以下は⑥にかかる。
* in relation to ～：〔熟語〕～に関して
* knowledge：〔名詞〕知識 → know〔動詞〕
* understanding：〔名詞〕理解
* proficiency：〔名詞〕熟達，熟練＝ skill：技能
* list：〔動詞〕列挙する
* listed：〔過去分詞〕列挙されている
  listed 以下は，前出の名詞 the knowledge, understanding and proficiency にかかる。
* under：〔前置詞〕（種類，分類など）…のなかに，…の項目下に
* maneuvering and handling：〔名詞〕両方とも「操船」という意味。 handle：〔他動詞〕操作する，操船する
* in column 2 of table A-Ⅱ/2―：表AのⅡ/2における2縦列目の

①②③④⑤⑥部分訳 _____

全文和訳 _____ 知らされているべきである。

6) Specification of the minimum standard of competence/ for masters and chief mates/ of ships of 500 gross tonnage or more/; and

* 文頭には It is（それは～である）が省略されている。It は 5) の column 2 of table A-Ⅱ/2―を指す。specification 以下は，column 2 of table A-Ⅱ/2―に書かれている具体例。―（ダッシュという）は「つまり」という意味である。
* specification：〔名詞〕詳細，明細，特記仕様書
* minimum standard of competence for ～：～のための権限に関する最低基準
* competence：〔名詞〕権限
* gross：〔形容詞〕総…，全体の
* tonnage：〔名詞〕船舶のトン数
* ships of 500 gross tonnage or more：総トン数が500トンかそれ以上の船舶

全文和訳 _____

7) (should) **be made** thoroughly familiar with the use of all navigational and maneuvering aids/ fitted in the ship concerned,/ including their capabilities and limitations./

* thoroughly：〔副詞〕十分に，徹底的に，完全に
* 主語は，4) の②と同じ masters and chief mates である。
* (masters and chief mates) should be made thoroughly familiar with ～：大意「船長と一等航海士は，～に十分に精通しているべきである。」 be made は使役動詞 make の受動態。
* be fitted in ～：〔熟語〕～に備え付けられている
* the ship concerned：当該の船，関係している船。concerned は ship にかかる形容詞。
* including：〔前置詞〕
* capability and limitation：性能と限界
* (including their capabilities and limitations) と (fitted in the ship concerned) は aids（援助装置）に意味がつながる。
* use of all navigational and maneuvering aids：航海用および操船用のすべての航法装置の使用
* manoeuvering はイギリス英語のスペル。アメリカ英語のスペルは maneuvering。辞書には両方記載されている。

全文和訳 _____

30

Step 3
網掛けは主語，**ゴシック体**は動詞（述語動詞）。音読が終わったら Step Check Box をぬりなさい。

Guidance/ regarding additional training/ for masters and chief mates/ of large ships and ships/ with unusual maneuvering characteristics/

1 It **is** important/ that masters and chief mates **should have had** relevant experience and training/ before assuming the duties/ of master or chief mate/ of large ships or ships/ having unusual maneuvering and handling characteristics/ significantly different from those/ in which they **have** recently **served**./ Such characteristics **will** generally **be found** in ships/ which **are** of considerable deadweight or length of special design/ or of high speed./

2 Prior to their appointment to such a ship,/ masters and chief mates **should**:

  (1) **be informed of** the ship's handling characteristics/ by the company,/ particularly in relation to the knowledge, understanding and proficiency/ listed under ship manoeuvering and handling/ in column 2 of table A-II/2— Specification of the minimum standard of competence/ for masters and chief mates/ of ships of 500 gross tonnage or more/; and

  (2) **be made** thoroughly familiar with the use of all navigational and manoeuvering aids/ fitted in the ship concerned,/ including their capabilities and limitations./

【専門英語単語リスト】書いておぼえよう

発音したり，スペル練習したり，意味を調べたりしたらチェック欄の○をぬりなさい。
辞書を引いて単語の意味が多くある場合，英語の文脈から判断し，適切な意味を選びなさい。

英文	単語	品詞	意味を調べて書きなさい	発音しながらスペル練習しなさい	チェック
1)	regarding	前置詞			○○○
1)	maneuver	他動詞			○○○
2)	relevant	形容詞			○○○
2)	assume	他動詞			○○○
3)	deadweight	名詞			○○○
4)	appointment	名詞			○○○
5)	be informed of ～	熟語			○○○
5)	in relation to ～	熟語			○○○
5)	proficiency	名詞			○○○
5)	list	他動詞			○○○
6)	specification	名詞			○○○
6)	competence	名詞			○○○
7)	be familiar with	熟語			○○○
7)	capability / capable	名／形			○○○
7)	limitation / limit	名／動			○○○

Unit 7　船員の資格証明

Step Check Box □ □ □

Step 1　下記の英文を読み，次の3つの活動をしなさい。終了後Step Check Box □をぬりなさい。

(1) 動詞（述語動詞）を○で囲む。動詞の意味を考えて，動詞のあとに続く内容を推測する。
(2) 主語（主部）に下線を引く。
(3) スラッシュを入れる。スラッシュを入れる場所は主に下記の3つ。
- コンマ(,)やピリオド(.)でスラッシュ(/)を入れる。音読するときは，スラッシュで息つぎする。
- 接続詞の前にスラッシュを入れる。［接続詞＋主語＋動詞］は文としてまとまった意味を表す。
  ここでいう接続詞とは when, if, until, before, after など（and, but, or, so を含まない）。
- 原則として前置詞の前にスラッシュを入れる。（前置詞＋名詞）は，最小の意味のまとまりを表す。

1) Ships, except those excluded by Article III, are subject, while in the ports of a Party, to control by officers duly authorized by that Party to verify that all seafarers serving on board who are required to be certificated by the Convention are so certificated or hold an appropriate dispensation.

2) Such certificates shall be accepted unless there are clear grounds for believing that a certificate has been fraudulently obtained or that the hold of a certificate is not the person to whom that certificate was originally issued.

3) When exercising control under this Article, all possible efforts shall be made to avoid a ship being unduly detained or delayed.

4) If a ship is so detained or delayed, it shall be entitled to compensation for any loss or damage resulting therefrom.

5) This Article shall be applied as may be necessary to ensure that no more favourable treatments is given to ships entitled to fly the flag of non-Party than is given to ships entitled to fly the flag of a Party.

出典：1級海技士（航海）平成26年4月定期試験問題
STCW条約 Article Xからの抜粋

Step 2　網掛けは主語，ゴシック体は動詞（述語動詞），＊語注・文法的留意事項，①②③は訳順番号

(1) 単語や熟語の説明を読みながら，部分訳を書きなさい。わからない単語の意味は調べなさい。
(2) 部分訳が終わったら，①②③の訳順番号を参考に，全文和訳を書きなさい。
　全文和訳が終わったら，Step Check Box □をぬりなさい。

1) Ships,／ except those (excluded by Article III),／ **are subject**,／
　　②　　　　　　　　①　　　　　　　　　　　　　⑫

* except：〔前置詞〕〜を除く
* excluded：〔過去分詞〕〜により除外された（船）
* be subject to control：〜の支配下にある，〜に左右される
* subject：〔形容詞〕〜を受けやすい，支配を受けて
* except those (excluded by Article III) は，shipsにかかる（修飾する）。
* those = ships
* exclude：〔他動詞〕を除く ↔ include〔他動詞〕を含む
* 英文の概要：②船は，③に停泊中は，④⑤⑥⑦⑧that以下を証明するために，⑩⑪⑫の航海士の支配下に置かれる

①②部分訳　_____

[while (they **are**) in the ports of a Party,]／ **to** control by officers／ duly authorized by that Party／
　　　　　③　　　　　　　　　　　　　　　⑫　　　　　　　　　⑪　　　　　　　⑩

* while ships are in the ports of a Party：締約国の港に停留中
* duly：〔副詞〕十分に，適切に，正しく
* authorized：〔過去分詞〕〜により権限を与えられた

⑩⑪⑫部分訳　_____

Unit 7 船員の資格証明

　　　　　　　　　⑨　　　　　　⑥　　　　　　　　　　　　　　　⑤　　　　　　　　　　　④
to verify/ [that all seafarers serving on board/ who **are required** to be certificated by the Convention/

* verify：〔他動詞〕〜を証明する　（何を証明するかは，that 以下に述べられている）
* all seafarers serving on board：乗船中のすべての船員は　（serving は現在分詞で seafarers にかかる）
* who：〔関係代名詞〕who 以下は，直前の名詞 all seafarers にかかる。
* be required：求められている　　　　　　　　　　　　* be certificated by：〜により証明書で認定されるように
* convention：〔名詞〕国際条約

④⑤⑥部分訳 _____

　　　　　　⑦　　　　　　　　　　　⑧
are so **certificated**/ or **hold** an appropriate dispensation.]/

* certificate：〔他動詞〕〜に証明書を与える　　　　* appropriate：〔形容詞〕適切な，妥当な
* dispensation：〔名詞〕（法令などの）適用免除　　* 主語④⑤⑥の動詞が⑦⑧である。

⑦⑧部分訳　そのような証明書を与えられているか，あるいは適切な法令適用免除を保持しているか

全文和訳

　　　　　　　　　　⑩　　　　　　　　　　⑪　　　　　　　　　⑨　　　　　⑧　　　　　　　⑦
2) Such certificates **shall be accepted**/ [unless there **are** clear grounds/ for believing/

* such certificates shall be accepted：そのような証明書は受託されるべきである
* unless：〔接続詞〕もし〜でないのならば = if not
* [unless there are ①②③④⑤⑥⑦⑧]が，接続詞 unless の意味が続く範囲
* clear grounds：はっきりとした理由
* for believing：believe の目的語は，下記の that 以下の文 2 つ「①②③ or ④⑤⑥」である。
　for が前置詞なので，believing は動名詞である。

⑦⑧⑨部分訳　that 以下を _____

　　　　①　　　　　　③　　　　　②　　　　　③
[that a certificate **has been** fraudulently **obtained**]/ or

* has been obtained：これまでに〜されている。現在完了形の受動態。　* fraudulently：〔副詞〕不正に
* obtain：〔他動詞〕取得する = get
* or：〔接続詞〕A（①②③ であるということ），あるいは B（④⑤⑥であること）

①②③部分訳 _____ ことを，あるいは

　　　　　　④　　　　　　　　　　　⑥　　　　　　　　　　　⑤
[that the hold of a certificate **is not** the person/ to whom that certificate **was** originally **issued**.]/

* the hold of a certificate：証明書の保有（所有）
* to whom that certificate was issued = that certificate was originally issued to the person：その証明書がもともと発行された人
* originally：〔副詞〕　　　　　　　　　　　　* original：〔形容詞〕原の，根源の，最初の

④⑤⑥部分訳　証明書の保有者が，_____ ということを

全文和訳

33

3) When exercising control under this Article,/
   ②　　　　　　　①

   * exercising 以下は分詞構文なので，下記のように書き換えることができる。
   = when you exercise control under this Article：この条項の下，統制を行使するときは

   ③　　　　　　　　⑦　　　　　　⑥　　　　　　　⑤　　　　　　④
   all possible efforts shall be made/ to avoid a ship being unduly detained or delayed./

   * make an effort：〔熟語〕努力をする→受動態にすると，all possible efforts shall be made になる。
   * avoid Ving（動名詞）：〔他動詞〕～することを避ける
   * being unduly detained or delayed の主語は a ship である。「船が～するのを避けるために」と訳す。
   * unduly：〔副詞〕不正に＝wrongly ⟷ duly：正しく，十分に，適切に
   * detain：〔他動詞〕～を拘留する　　　　　　　　　* delay：〔他動詞〕～を遅延させる

   ④⑤⑥部分訳　_____　が不正に拘留させられたりあるいは遅延させられるのを　_____

   全文和訳　_____

4) [If a ship is so detained or delayed,]/
                    ①

   ①部分訳　もし，船舶がそのように　_____

   ②　　　⑥　　　　⑤　　　　　　　　④　　　　　　　③
   it shall be entitled to compensation/ for any loss or damage/ resulting therefrom./

   * be entitled to 名詞：～する権利が与えられる
   * compensation：〔名詞〕補償，弁償　　　　　　　　* compensate：〔動詞〕
   * any loss or damage：いかなる損失や被害に対する
   * resulting therefrom：現在分詞 resulting が④にかかる。「そこから結果として生じる」と訳す。

   全文和訳　_____

5) This Article shall be applied/ [as (it) may be necessary]/ to ensure/
   ⑩　　　　　⑪　　　　　　　　　⑨　　　　　　　　　　　　⑧

   * shall be applied：適用されるべきである
   * as (it) may be necessary：必要な場合は。it は省略されている。
   * to ensure that：that 以下を確認するために

   ⑩⑪部分訳　_____

   ⑥　　　　　　　　　　　　⑦　　　　　⑤　　　④
   [that no more favourable treatment is given to ships/ entitled to fly the flag of non-Party/

   * no more favourable treatment：より好都合な処遇が　*主語が否定なので動詞を否定にして訳す。
   * entitled to fly the flag of non-Party：過去分詞 entitled は ships にかかる。「～する権利を与えられる船」となる。

   ④⑤⑥⑦部分訳　締約国ではない国を旗国とする船に，都合のよい処遇が与えられないように

   ③　　②　　　　　①
   than is given to ships/ entitled to fly the flag of a Party.]/

   ①②③部分訳　締約国を旗国とする船に与えるより

   全文和訳　_____

Step 3   網掛け は主語，**ゴシック体**は動詞（述語動詞）

(1) 音声を聞きながら，音読する。音読が終わったら，Step Check Box □をぬりなさい。
(2) 左から順に，スラッシュ毎の意味を考える。

Ships,/ except those excluded by Article Ⅲ,/ **are subject**,/ while in the ports of a Party,/ **to** control by officers/ duly authorized by that Party/ to verify that all seafarers serving on board/ who **are required** to be certificated by the Convention/ **are** so **certificated**/ or **hold** an appropriate dispensation./ Such certificates **shall be accepted**/ unless there **are** clear grounds/ for believing that a certificate **has been** fraudulently **obtained**/ or that the hold of a certificate **is not** the person/ to whom that certificate **was** originally **issued**./

When exercising control under this Article,/ all possible efforts **shall be made**/ to avoid a ship being unduly detained or delayed./ If a ship **is** so **detained** or **delayed**,/ it **shall be entitled to** compensation/ for any loss or damage/ resulting therefrom./

This Article **shall be applied**/ as may be necessary/ to ensure/ that no more favourable treatment **is given** to ships/ entitled to fly the flag of non-Party/ than **is given** to ships/ entitled to fly the flag of a Party./

【専門英語単語リスト】書いておぼえよう

発音したり，スペル練習したり，意味を調べたりしたらチェック欄の○をぬりなさい。
辞書を引いて単語の意味が多くある場合，英語の文脈から判断し，適切な意味を選びなさい。

英文	単語	品詞	意味を調べて書きなさい	発音しながらスペル練習しなさい	チェック
1)	exclude / include	他動詞			○○○
1)	be subject to 名詞	熟語			○○○
1)	authorize / authority	動 / 名			○○○
1)	verify	他動詞			○○○
1)	certificate / certification	動 / 名			○○○
2)	accept / acceptance	動 / 名			○○○
2)	unless	接続詞			○○○
2)	grounds (しばしば複数形で)	名詞			○○○
2)	obtain	他動詞			○○○
3)	avoid + Ving	他動詞			○○○
3)	unduly / duly	副詞			○○○
3)	detain	他動詞			○○○
3)	delay	他動詞			○○○
4)	compensate / compensation	名 / 動			○○○
4)	loss / lose	名 / 動			○○○

Unit 8 沿岸航海に従事する船員の要件　Step Check Box □ □ □

Step 1　下記の英文を読み，次の3つの活動をしなさい。終了後 Step Check Box □ をぬりなさい。

(1) 動詞（述語動詞）を○で囲む。動詞の意味を考えて，動詞のあとに続く内容を推測する。
(2) 主語（主部）に下線を引く。
(3) スラッシュを入れる。スラッシュを入れる場所は主に下記の3つ。
  - コンマ (,) やピリオド (.) でスラッシュ (/) を入れる。音読するときは，スラッシュで息つぎする。
  - 接続詞の前にスラッシュを入れる。[接続詞＋主語＋動詞] は文としてまとまった意味を表す。
    ここでいう接続詞とは when, if, until, before, after など (and, but, or, so を含まない)。
  - 原則として前置詞の前にスラッシュを入れる。（前置詞＋名詞）は，最小の意味のまとまりを表す。

1) Principles governing near-coastal voyages
2) Any Party defining near-coastal voyages for the purpose of the Convention shall not impose training, experience or certification requirements on the seafarers serving on board the ships entitled to fly the flag of another Party and engaged on such voyages in a manner resulting in more stringent requirements for such seafarers than for seafarers serving on board ships entitled to fly its own flag.
3) In no case shall any such Party impose requirements in respect of seafarers serving on board ships entitled to fly its own flag.
4) Seafarers serving on a ship which extends its voyage beyond what is defined as a near-coastal voyage by a Party and enters waters not covered by that definition shall fulfill the appropriate competency requirements of the Convention.

（注）　Party：締約国（条約や契約を結んでいる国）

出典：1級海技士（航海）平成25年10月定期試験問題
STCW 条約附属書 Regulation 1/3 からの抜粋

Step 2　網掛け は主語，**ゴシック体**は動詞（述語動詞），＊語注・文法的留意事項，①②③は訳順番号

(1) 単語や熟語の説明を読みながら，部分訳を書きなさい。わからない単語の意味は調べなさい。
(2) 部分訳が終わったら，①②③の訳順番号を参考に，全文和訳を書きなさい。
    全文和訳が終わったら，Step Check Box □ をぬりなさい。

　　　　　②　　　　①
1) Principles/ governing near-coastal voyages/

  * govern：〔他動詞〕管理する，統治する
  * Ving の働きは2つ。現在分詞か動名詞のどちらかである。動名詞は名詞として働き，主語になる。
    現在分詞は形容詞として働き，名詞を修飾し「〜している」と訳す。governing は principles にかかるので，現在分詞の働きをしている。
  * principle：〔名詞〕原則，原理　　　　　　＊ near-coastal voyage：沿岸航海

タイトル和訳
_____

　　　　　　③　　　　　②　　　　　　　　　　　　①　　　　　　　　　　⑬
2) Any Party /defining near-coastal voyages/ for the purpose of the Convention **shall not impose**

  * define：〔動詞〕定義する，境界を定める　　＊ defining：〔現在分詞〕直前の名詞 party にかかる。
  * purpose：〔名詞〕趣旨，意図，目的
  * convention：〔名詞〕国際条約，協約。treaty よりくだけた表現。
  * impose A on 人：人に A（義務，仕事，罰金，税など）を課す
    動詞 impose の後に，どういう内容が続くか？
    この動詞は他動詞なので，この動詞 impose の後に「〜を」（名詞）という言葉が続く。
  * 動詞が自動詞か他動詞か？（主語 S ＋動詞 V）の後に，目的語（〜を）として働く名詞が続く場合，その動詞は他動詞として機能する。

36

①②③部分訳＿＿＿＿＿＿＿＿＿＿＿＿＿＿＿＿＿＿＿＿＿＿＿＿＿＿＿＿＿＿＿＿＿＿＿＿

training, experience or certification requirements/ **on** the seafarers serving on board the ships/ entitled to fly the flag of another Party/
（⑫　　　　　　　　　　　　　　　　　　⑪　　　　　　　　　　⑩　　　　　　　　⑨）

- certification requirement：必要条件の資格
- serving：Ving の前に名詞 seafarers があるので，serving は現在分詞で，seafarers にかかる。
- entitled：〔過去分詞〕〜の資格が与えられて。直前の名詞 ships にかかる。
  shall not impose が動詞なので，entitled や engaged は過去形ではなく，過去分詞形である。
- ⑨の意味：他の締約国を旗国とする（資格がある）。「資格がある」はとくに訳さなくてもよい。

and engaged on such voyages in a manner/ resulting in more stringent requirements for such seafarers/
（⑧　　　　　　　　　　　　　　　　　　　　　　　　⑦）

- engaged on：〔過去分詞〕〜に従事している，携わっている（船員）。⑪の the seafares にかかる。
- in a manner resulting in 〜：resulting は現在分詞形で，resulting in 以下 its own flag までの文は manner にかかる。
- more stringent requirements than…：stringent の比較級
- stringent：厳しい＝ strict
- ⑦⑧大意：結果的に than 以下④⑤の船員より厳しい資格を（船員に）求めるようなやり方で航海に従事して

⑨⑩部分訳＿＿＿＿＿＿＿＿＿＿＿＿＿＿＿＿＿＿＿＿＿船舶に乗船して従事している船員に

than for seafarers serving on board ships/ entitled to fly its own flag./
（⑥　　　　⑤　　　　　　　　　　　　　　④）

- ④⑤⑥大意：自国を旗国とする船舶に乗船して従事している船員より

全文和訳＿＿

3) In no case/ **shall** any such Party **impose** requirements/ in respect of seafarers serving on board ships/ entitled to fly its own flag./
（①　　　⑤　　　　　　　　⑦　　　　　　　⑥　　　　　④　　　　　　　　　③　　　　②）

- In no case：文頭が否定語（no case）なので，主語 S（any such Party）と動詞 V（shall impose）が疑問文の語順になっている。これを倒置文という。no case と名詞が打ち消しなので，訳すときは，動詞を打ち消す。
- ①⑤⑦大意：いかなる場合もそのような締約国は〜を課してはいけない
- in respect of：〔熟語〕〜に関しては（＝ with respect to）
- its own flag：自国の旗

全文和訳＿＿

4) Seafarers serving on a ship/ [which **extends** its voyage/ beyond what is defined/ as a near-coastal voyage
（⑥　　　⑤　　　　　　　　　　　　②　　　　　　　①　　　　　　　　①）

- which：〔関係代名詞〕先行詞は ship。[which 以下，次の行 by that definition まで]が ship を修飾する。
- extend：〔他動詞〕〜を広げる
- beyond：〔前置詞〕〜を超えて
- what is defined as a near-coastal voyage by a Party：締約国により沿岸航海として境界を定めている範囲
- what：〔関係代名詞〕＝ the things which の意味で，先行詞が the things。what は「〜のもの」と訳す。
- ⑤⑥が主語，⑨ shall fulfill が動詞である。
- ②大意：（〜の範囲以上に）航海を広げる（船に乗って従事する船員）

①部分訳＿＿

① ④ ③ ⑨
by a Party/ and (which) **enters** waters/ not covered by that definition] **shall fulfill**
⑧ ⑦
the appropriate competency requirements of the Convention./

* waters：〔名詞〕海域
* ③④⑤⑥大意：その境界により保護されていない海域に入る（船に乗って従事する船員）
* covered：〔過去分詞〕動詞が shall fulfill だから，covered は過去分詞で，waters にかかる。
* shall fulfill：この主語は，⑥Seafares である。助動詞は，（述語）動詞を探すときの目印になる。
* fulfill：〔他動詞〕（条件，要求，義務などを）満たす，果たす ＝ carry out, complete
* appropriate：〔形容詞〕適切な，妥当な
* competency：（competence が普通）〔名詞〕能力，適性，法的資格，法的権限，法的行為能力
* the appropriate competency requirements：適切な能力要件

全文和訳

Step 3   網掛け は主語，**ゴシック体**は動詞（述語動詞）

(1) 音声を聞きながら，音読する。音読が終わったら，Step Check Box □をぬりなさい。
(2) 左から順に，スラッシュ毎の意味を考える。

Principles/ governing near-coastal voyages/
　Any Party/ defining near-coastal voyages/ for the purpose of the Convention **shall not impose** training, experience or certification requirements/ **on** the seafarers serving on board the ships/ entitled to fly the flag of another Party/ and engaged on such voyages in a manner/ resulting in more stringent requirements for such seafarers/ than for seafarers serving on board ships/ entitled to fly its own flag./
　In no case/ **shall** any such Party **impose** requirements/ in respect of seafarers serving on board ships/ entitled to fly its own flag./
　Seafarers serving on a ship/ which **extends** its voyage/ beyond what is defined/ as a near-coastal voyage by a Party/ and **enters** waters/ not covered by that definition **shall fulfill** the appropriate competency requirements of the Convention./

【文法のポイント】
接続詞の前，原則として前置詞の前で，スラッシュを入れる。スラッシュは，意味のまとまりを示す。スラッシュ毎に，左から右へ区切りながら訳せば，英文の内容が理解できる。和訳をする場合は，下記のポイントを参考に，日本語らしい語順に直す。

1. タイトルは，次に続く英文内容と密接につながる。ここでは「沿岸航海を規制する指針（原則）」を専門知識として知っていることが重要である。
2. 1級海技士試験問題は，1つの英文が長い。ピリオド，コンマ，接続詞や前置詞は，意味の切れ目（どこからどこまでが意味のまとまりなのか）を示してくれる。スラッシュを入れるとわかりやすい。
3. 助動詞は，述語動詞を探すときの目印になる。
4. （前置詞＋名詞）は，直前（or 直後）の名詞あるいは動詞にかかる。
5. ［接続詞 because, when, if, while ＋主語 S ＋動詞 V］は，主語 S ＋動詞 V の動詞にかかる
6. ［関係代名詞 which/that ＋主語 S ＋動詞 V］［関係副詞 where, when ＋主語 S ＋動詞 V］は直前の名詞にかかる。
7. ただし先行詞のない関係代名詞 what は「〜のもの」と訳す。［what is defined as a near-coastal voyage by a Party］→「締約国により沿岸航海として定義されているもの」という意味になる。ここでは，what を「なに？」（疑問詞）とは訳さない。

【専門英語単語リスト】書いておぼえよう

発音したり，スペル練習したり，意味を調べたりしたらチェック欄の○をぬりなさい。
辞書を引いて単語の意味が多くある場合，英語の文脈から判断し，適切な意味を選びなさい。

英文	単語	品詞	意味を調べて書きなさい	発音しながらスペル練習しなさい	チェック
1)	principle	名詞			○○○
1)	govern / government	動 / 名			○○○
1)	coastal / coast	形 / 名			○○○
2)	define / definition	動 / 名			○○○
2)	impose	他動詞			○○○
2)	certification / certificate	名詞			○○○
2)	entitle	他動詞			○○○
2)	result in / result	動 / 名			○○○
2)	stringent	形容詞			○○○
3)	in respect of	熟語			○○○
4)	beyond	前置詞			○○○
4)	fulfill	他動詞			○○○
4)	appropriate	形容詞			○○○
4)	competence	名詞			○○○
4)	convention / treaty	名詞			○○○

Unit 9　遭難信号

Step Check Box □ □ □

Step 1　下記の英文を読み，次の３つの活動をしなさい。終了後 Step Check Box □ をぬりなさい。

(1) 動詞（述語動詞）を○で囲む。動詞の意味を考えて，動詞のあとに続く内容を推測する。
(2) 主語（主部）に下線を引く。
(3) スラッシュを入れる。スラッシュを入れる場所は主に下記の３つ。
  - コンマ (,) やピリオド (.) でスラッシュ (/) を入れる。音読するときは，スラッシュで息つぎする。
  - 接続詞の前にスラッシュを入れる。［接続詞＋主語＋動詞］は文としてまとまった意味を表す。
    ここでいう接続詞とは when, if, until, before, after など (and, but, or, so を含まない)。
  - 原則として前置詞の前にスラッシュを入れる。（前置詞＋名詞）は，最小の意味のまとまりを表す。

1) It is important that all means for indicating the position of ships in distress or survival craft should be properly used.

2) Radio transmissions should be made as soon as possible but other means, e.g. rockets and hand flares, should be conserved until it is known that they may attract the attention of aircraft or ships in the vicinity.

3) The attention of masters is directed/ to the great advantage of prior training, so that as many of the ship's crew as possible are familiar with the proper use of all of the appliances provided for their safety.

出典：2級海技士（航海）平成20年7月定期試験問題

Step 2　網掛けは主語，ゴシック体は動詞（述語動詞），＊語注・文法的留意事項，①②③は訳順番号

(1) 単語や熟語の説明を読みながら，部分訳を書きなさい。わからない単語の意味は調べなさい。
(2) 部分訳が終わったら，①②③の訳順番号を参考に，全文和訳を書きなさい。
  全文和訳が終わったら，Step Check Box □ をぬりなさい。

1) It **is** important　⑥

 * it：仮主語なので，「それは」と訳さない。that 以下全部が真の主語なので，「〜ということは」と訳す。
   ここでは，［that all means for… should be properly used］が真の主語である。

　　　　④　　　　　③　　　　　　　　②　　　　　　　　　①
[that all means/ for indicating the position/ of ships in distress or survival craft

 * that：〔接続詞〕It is important と that 以下とをつなぐ働きをする。
   that 以下には必ず主語と動詞がある。主語 S は all means…survival craft で，動詞 V は should be properly used
 * ships in distress：遭難船舶　　　　＊ survival craft：救命艇
 * ①②③④の意味がわかると，全体で何が問われているか推測できる。遭難船舶に関する専門知識が必須である。

①②③④部分訳 _____

　　　　　　　　　　⑤
should be properly **used**.]/

 * be + Ved：受動態（受身）は頻繁に使われる。「〜られる」と訳す。
 * should：〔助動詞〕shall や should は法令文のなかでよく使われる。「〜すべきである」と訳す。助動詞は（述語）動詞の一部と考え，動詞を○で囲む場合は，助動詞も○で囲む。

全文和訳 _____

　　　　　　　　　①　　　　　　　　　③　　　　　　②
2) Radio transmissions **should be made**/ as soon as possible/

 * radio transmission：無線通信　　　　　　　　　　＊ should be made：〔受動態〕（無線通信は）行われるべきである
 * as 副詞 as possible：〔熟語〕できるだけ…に　　　＊ as soon as possible：できるだけ早く

40

①②③部分訳 _____

but other means,/ e.g. rockets and hand flares,/ **should be conserved**/
　　⑤　　　　　　　④　　　　　　　　　　　　　⑬

* other means：他の手段は　　　　　　　　　　　＊ e.g.：たとえば，for example と読む。
* conserve：〔他動詞〕保存する，節約する　　＊ should be conserved：〔受動態〕節約されるべき(=使用を控えるべき)である
* but：〔接続詞〕but があると，前に出てきた情報を否定する内容が，後に続くということがわかる。

④⑤部分訳 _____

[until **it is known**/ that they **may attract** the attention/ of aircraft or ships/ in the vicinity.]/
　⑫　⑪　　　　　⑥　　　　　　　⑩　　　　　⑨　　　　　　⑧　　　　　　⑦

* until：〔接続詞〕〜するまで　　＊接続詞 until の後には必ず主語(it)と動詞(is known)がくる。
* it：仮主語の it なので「それは」とは訳さない。ここでは that 以下全文(⑥〜⑩)が真の主語である。
* that：〔接続詞〕後には必ず主語 S と動詞 V がくる。ここでは，S は they，V は may attract。
* in the vicinity は前の名詞 aircraft or ships にかかる。「付近にいる航空機や船舶」となる。

⑥⑦⑧⑨⑩部分訳 _____ということが

* この文の主語は Radio transmissions で，動詞は should be made である。だから主語①以下の文には，主語①「無線通信」と動詞③「行われる」に関係する詳細(具体的)な情報が述べられる。
* until it is known that 以下の文は，主語⑥と動詞⑩だけでは，何をどう引きつけるのかわからない。そこで，⑥⑩の後には，「何を引きつけるのか」という詳細(具体的)な情報が書いてある。
* 専門科目の知識があれば，主語①と動詞③さえわかれば，書かれている内容を推測することができる。

全文和訳
_____

3) The attention of masters **is directed**/ to the great advantage of prior training,/
　　　⑦　　　　　　　　　　　　　⑨　　　　　　　　⑧

* is directed：〔受動態〕〜に向けられる　　　　＊ prior training：事前訓練
* to the great advantage of prior training：「事前訓練の大いなる利点」→「事前訓練がとても有利であるということ」と，わかりやすく言い直す。和訳するときは，むずかしい表現をわかりやすい語句で言い換えるスキルが必要である。

⑦⑧⑨部分訳 _____

[so that as many of the ship's crew as possible **are familiar with** the proper use/ of all the appliances/ provided
　⑥　　　　　①　　　　　　　　　　　　　⑤　　　　　　　④　　　　　　　③　　　　　　②

for their safety.]/

* so that = in order that：〔接続詞〕〜するように，〜するために。so that の後に，主語 S①+動詞 V⑤が続く。
* as many of the ship's crew as possible：as 〜 as possible の形。「できるだけ多くの船員が」という意味。
* are familiar with 〜：この語句の後に，何に精通しているのかという具体的な情報(④③②)が述べられる。
* appliances provided for their safety：安全のために用意されている設備
* 名詞③+過去分詞 provided：過去分詞の直前に名詞が付くと，過去分詞は直前の名詞③にかかる。

④③②部分訳 _____

全文和訳
_____

Step 3　網掛けは主語，**ゴシック体**は動詞（述語動詞）

(1) 音声を聞きながら，音読する。音読が終わったら，Step Check Box □をぬりなさい。
(2) 左から順に，スラッシュ毎の意味を考える。

It **is** important that all means/ for indicating the position/ of ships in distress or survival craft **should be** properly **used**./ Radio transmissions **should be made**/ as soon as possible/ but other means,/ e.g. rockets and hand flares,/ **should be conserved**/ until it **is known**/ that they **may attract** the attention/ of aircraft or ships/ in the vicinity./ The attention of masters **is directed**/ to the great advantage of prior training,/ so that as many of the ship's crew as possible **are familiar with** the proper use/ of all of the appliances/ provided for their safety./

【文法のポイント】
接続詞の前，前置詞の前で，スラッシュを入れる。スラッシュは，意味のまとまりを示す。スラッシュ毎に，左から右へ区切りながら訳せば，英文の内容が理解できる。和訳をする場合は，下記のポイントを参考に，日本語らしい語順に直す。

1. ［接続詞＋主語S＋動詞V…］は，動詞Vにかかる。
2. （前置詞＋名詞（もの・コト））は，直前（直後）の名詞や動詞にかかる。
3. ［関係代名詞＋主語S＋動詞V…］は，直前の名詞にかかる。
4. 名詞＋過去分詞(Ved)の語順の場合，過去分詞(Ved)は，直前の名詞にかかる。
5. 名詞＋現在分詞(Ving)の語順の場合，現在分詞(Ving)は，直前の名詞にかかる。

【専門英語単語リスト】書いておぼえよう

発音したり，スペル練習したり，意味を調べたりしたらチェック欄の○をぬりなさい。
辞書を引いて単語の意味が多くある場合，英語の文脈から判断し，適切な意味を選びなさい。

英文	単語	品詞	意味を調べて書きなさい	発音しながらスペル練習しなさい	チェック
1)	means	名詞			○○○
1)	indicate	他動詞			○○○
1)	distress	名詞			○○○
1)	survival	名詞			○○○
2)	transmission	名詞			○○○
2)	hand flare	名詞			○○○
2)	conserve	他動詞			○○○
2)	attract	他動詞			○○○
2)	attention	名詞			○○○
2)	aircraft	名詞			○○○
3)	direct	他動詞			○○○
3)	advantage	名詞			○○○
3)	be familiar with	熟語			○○○
3)	appliance	名詞			○○○
3)	provide	他動詞			○○○

理解を深めるキーワード「Rocket と Hand Flare」

　遭難の際，他船や陸上の機関などへ，遭難した位置をいち早く知らせ，救援を呼ぶことが重要となる。しかし広い大海原において，気象海象状況などによって，船の位置をおおよそ把握できていても，詳しい位置を特定できないことがある。そこで，あらゆる手段（電波，視覚，聴覚，嗅覚など）によって，他船などへ自分の船が遭難していることを伝えられるように，遭難信号用の道具を船舶に積み込み，航海することが国際的なルールで定められている。

　そのなかに昼夜問わずに使える，花火のような，光や煙によって信号を送るものがある。その代表的なものとして，"Rocket"（Rocket Parachute signal（落下傘付信号），Rocket Star signal（火せん））（図参照）と呼ばれる，300mほど上空で赤色の閃光を3秒以上発するもの，Hand Flare（手持ち信号紅炎）と呼ばれる，手に持って，赤色の花火のような信号を発するものがある。これらは，遠くまで，長い時間，相手に伝達することができないため，相手が付近にいると推測できるときにのみ使用する。

Rocket（Rocket Parachute signal：落下傘付信号）

Unit 10　海難救助と責任

Step Check Box □ □ □

Step 1　下記の英文を読み，次の3つの活動をしなさい。終了後 Step Check Box □をぬりなさい。

(1) 動詞（述語動詞）を○で囲む。動詞の意味を考えて，動詞のあとに続く内容を推測する。
(2) 主語（主部）に下線を引く。
(3) スラッシュを入れる。スラッシュを入れる場所は主に下記の3つ。
- コンマ (,) やピリオド (.) でスラッシュ (/) を入れる。音読するときは，スラッシュで息つぎする。
- 接続詞の前にスラッシュを入れる。［接続詞＋主語＋動詞］は文としてまとまった意味を表す。
  ここでいう接続詞とは when, if, until, before, after など (and, but, or, so を含まない)。
- 原則として前置詞の前にスラッシュを入れる。（前置詞＋名詞）は，最小の意味のまとまりを表す。

1) When a vessel suffers a casualty, or is otherwise in a position of peril, the master must decide as a matter of urgency where assistance, including salvage assistance, is needed or if the situation can be handled using the ship's own resources.

2) The authority of the master is not altered by engaging salvors.

3) He remains in command of the vessel despite the presence of a salvage master and he should therefore ensure that he is fully aware of the action taken in the rendering of salvage services.

4) Even though services have been accepted and assistance is being rendered, the salvor must cease his services if requested to do so by the master.

5) The master should, however, co-operate fully with the salvors, who are experts in salvage operations and should take account of any advice given by the salvage master or other person in charge of rendering or advising on salvage services.

出典：2級海技士（航海）平成20年10月定期試験問題

Step 2　網掛けは主語，ゴシック体は動詞（述語動詞），＊語注・文法的留意事項，①②③は訳順番号

(1) 単語や熟語の説明を読みながら，部分訳を書きなさい。わからない単語の意味は調べなさい。
(2) 部分訳が終わったら，①②③の訳順番号を参考に，全文和訳を書きなさい。
　　全文和訳が終わったら，Step Check Box □をぬりなさい。

1) [When a vessel **suffers** a casualty,/ or (when a vessel) **is** otherwise in a position of peril,]/
　　　　　　　①　　　　　　　　　　　　　　　　　　　　　　　②

- ＊ suffer：〔他動詞〕損害などを被る　　　　　　　　＊ casualty：〔名詞〕惨事，大事故
- ＊ ① or ②：or は接続詞で，ここでは「船が①のとき，あるいは②のとき」の意味。
- ＊ otherwise：別の状況で＝大事故を受けるという状況でなくて
- ＊ peril：〔名詞〕危険　　　　　　　　　　　　　　　＊ position：〔名詞〕情勢，形勢，位置，場所
- ＊ ①の主語が a vessel，動詞が suffers だから，①以下の英文では，船舶が被害を被ったことに関する詳細（具体的）な情報が述べられていることがわかる。専門知識があれば，主語と動詞がわかると，その後に続く内容を推測することができる。

①②部分訳 _____

the master **must decide**/ as a matter of urgency/
　　③　　　　　⑬　　　　　　　　　⑫

- ＊ as a matter of urgency：緊急のこととして
- ＊ decide：何を決定するのか，詳細（具体的）な情報が，次の2行に書いてある。つまり船長は「④⑤⑥⑦なのかあるいは⑨⑩⑪なのかどうかを」決定しなければいけない。

③⑫⑬部分訳 _____

[where assistance,/ including salvage assistance,/ **is needed**]/ or
　⑥　　　⑤　　　　　　　　　④　　　　　　　　　　⑦　　⑧

- ＊ where：〔疑問詞〕どこで　　　　　　　　　　　＊ assistance：〔名詞〕救助，支援

44

Unit 10 海難救助と責任

* salvage：〔名詞〕海難船舶の救助（沈没船の引き揚げ，座礁船の引き出し，航行不能船舶の曳航など），サルベージ
* including：〔前置詞〕～を含めて　　　　　　　　＊前置詞の後には名詞がくる。
* including salvage assistance：海難救助を含めて　　＊or：〔接続詞〕あるいは

④⑤⑥⑦⑧部分訳　どこで _____ あるいは

⑪　　　⑨　　　　　　⑪　　　　　　　　　　⑩
[if the situation **can be handled**/ using the ship's own resources.]/

* if：〔接続詞〕～かどうかを。「if 以下かどうかを，船長は decide しなければいけない」となる。「もし～ならば」という意味ではない。ここでは if 以下は動詞 decide の目的語なので「～かどうかを」と訳す。
* situation：〔名詞〕状況
* can be handled：can + 受動態の形　　　　　　　＊handle：〔他動詞〕～に対処する，～を解決する
* using：〔分詞構文〕using は動詞の意味を持ちながら，接続詞の働きをする。
* using the ship's own resources：船自身が持っている資源を使って→「自力で」とわかりやすく言い換える。

全文和訳
_____

　　　　　　　　　　　　　①　　　　　③　　　　　　　②
2) The authority of the master **is not altered**/ by engaging salvors./

* authority：〔名詞〕権限，機関
* is not altered：〔受動態〕変更されない　　　　　　＊alter：〔他動詞〕= change
* salvor / salver：〔名詞〕救助船
* engage：〔他動詞〕を雇う，引き込む = hire　　　　＊by engaging salvors：救助船を雇うことにより

全文和訳
_____

　　①　④　　　　　　③　　　　　　　　　　　　　②
3) He **remains** in command of the vessel/ despite the presence of a salvage master/ and

* remain：〔自動詞〕～がとどまる，のままである　　＊command：〔名詞〕指揮権，指令，号令
* despite：〔前置詞〕堅い表現で，in spite of（～にもかかわらず）と同じ意味。
* the presence of a salvage master：「サルベージマスター（海難救助の責任者）の存在」という意味なので，わかりやすいように「サルベージマスターがいる（にもかかわらず）」と訳す。

①②③④部分訳 _____ そして

⑤　　　　　　⑨
he **should** therefore **ensure**/

* therefore：〔副詞〕それによって，したがって　　　＊ensure：〔他動詞〕～を確認する，保証する
* should therefore ensure：したがって確認すべきである　＊確認する内容が [that 以下⑥⑦⑧] に詳細に述べられる。

　　　　⑧　　　　　　　⑦　　　　　　　　⑥
[that he **is** fully **aware of** the action/ taken in the rendering of salvage services.]/

* is fully aware of：～を十分に知っている
　⑥⑦では，何を知っているのか詳細な情報が述べられる。
* the action taken in ～：過去分詞 taken（受身の意味）が直前の名詞 the action を修飾している。
* the action taken in the rendering of salvage services：救助作業をしているときに取られる行為
* rendering：〔動名詞〕（援助・奉仕）をすること　　＊render：〔他動詞〕

⑥⑦⑧部分訳　自分は _____ ということを

全文和訳
_____

45

4) [Even though services **have been accepted**/ and assistance **is being rendered**,]/
   ⑤　　　　①　　　　　　②　　　　　　　③　　　　　　④

* even though = even if〔接続詞〕たとえ〜としても　　* Even though の後には主語と動詞が続く。
* service：〔名詞〕業務，仕事　　* assistance：〔名詞〕救助
* have been accepted：〔現在完了形受動態〕（支援業務が）過去に承諾依頼がなされ，現在，承諾業務は完了しているというイメージであるが，わかりやすく「承諾された」と訳す。
* is being rendered：〔現在進行形受動態〕いま（救助活動が）行われている

the salvor **must cease** his services/ [if (he **is**) requested to do so/ by the master.]/
　⑥　　⑩　　　　⑨　　　　　　　⑧　　　　　　　　　⑦

* salvor：〔名詞〕　　* cease：〔他動詞〕（仕事）をやめる
* is requested to do so by the master：船長により，そうするように求められる
* if he is requested to do so by the master：「もし〜ならば」の意味で，動詞 must cease にかかる。

⑦⑧部分訳 _____　⑥⑨⑩部分訳 _____

全文和訳 _____

5) The master **should**, however, **co-operate** fully with the salvors,/ [who **are** experts in salvage operations]/
   　②　　　　　①　　　　　　　　④　　　　　　③　　　　　　　　　　　　⑤

* should, however, co-operate fully with：しかしながら with 以下と十分に協力しなければならない
* salvors, who：関係代名詞 who 以下は，直前の名詞 salvors を修飾するが，who の前に（,）が付いているので，with the salvors,でいったん文を切り，「who（その救助者）は⑤だからである」と訳す。
* salvage operations：救助作業　　* be experts in：〜において専門家である

②③④⑤部分訳 _____

また　　　　⑥　　　　　⑫　　　　　　　⑪　　　　　　　⑦　　　　　　あるいは　⑩
and (the master) **should take account of** any advice/ given by the salvage master or other person/

* should take account of 〜：〜を考慮しなければならない
* any advice given by 〜：〜によって与えられる助言。過去分詞 given が直前の名詞 any advice を修飾する。

　　　　　　⑧　　　　　　　　　　⑨
in charge of rendering or advising on salvage services./

* in charge of 〜：〜に責任を持っている（サルベージマスターあるいは他の人）。⑧⑨は⑩を修飾している。
* rendering or advising on salvage services：海難救助業務を行っているか，あるいは助言を与える

⑦⑧⑨⑩部分訳 _____ によって与えられる

全文和訳 _____

Step 3　網掛けは主語，**ゴシック体**は動詞（述語動詞）

(1) 音声を聞きながら，音読する。音読が終わったら，Step Check Box □をぬりなさい。
(2) 左から順に，スラッシュ毎の意味を考える。

　When a vessel **suffers** a casualty,/ or **is** otherwise in a position of peril,/ the master **must decide**/ as a matter of urgency/ where assistance,/ including salvage assistance,/ **is needed**/ or if the situation **can be handled**/ using the ship's own resources./
　The authority of the master **is not altered**/ by engaging salvors./ He **remains** in command of the vessel/ despite the presence of a salvage master/ and he **should** therefore **ensure**/ that he **is** fully **aware of** the action/ taken in the rendering of salvage services./ Even though services **have been accepted**/ and assistance **is being rendered**,/

the salvor **must cease** his services/ if **requested** to do so/ by the master./
　The master **should**, however, **co-operate** fully **with** the salvors,/ **who are** experts in salvage operations/ and **should take account of** any advice/ given by the salvage master or other person/ in charge of rendering or advising on salvage services./

【専門英語単語リスト】書いておぼえよう

発音したり，スペル練習したり，意味を調べたりしたらチェック欄の○をぬりなさい。
辞書を引いて単語の意味が多くある場合，英語の文脈から判断し，適切な意味を選びなさい。

英文	単語	品詞	意味を調べて書きなさい	発音しながらスペル練習しなさい	チェック
1)	suffer	他動詞			○○○
1)	position	名詞			○○○
1)	decide	他動詞			○○○
1)	include	他動詞			○○○
1)	handle	他動詞			○○○
2)	engage	他動詞			○○○
3)	remain	自動詞			○○○
3)	in command of	熟語			○○○
3)	ensure	他動詞			○○○
3)	salvage	名詞			○○○
4)	accept	他動詞			○○○
4)	salvor / salver	名詞			○○○
5)	expert	名詞			○○○
5)	take account of	熟語			○○○
5)	in charge of	熟語			○○○

〜 理解を深めるキーワード「サルベージ」 〜

　海難により自船が自身で操船できない状況になったときに，救助機関（日本であれば海上保安庁）と「サルベージ」会社の部隊に救助を依頼する。サルベージ会社の救助部隊は，人命の救出援助だけではなく，船から原油などの液体が流失している場合，その流失を抑える任務や，転覆した船を元の状態に戻し，岸に船を運ぶ（曳航する）任務，また沈没した船を図に示すような大型の作業台船を利用し，引き上げる任務を行う。潜水作業が多く，さらに海難海域は気象海象が荒れている場合も多いため，任務を遂行するためには多くの経験と特殊技能を必要とする。

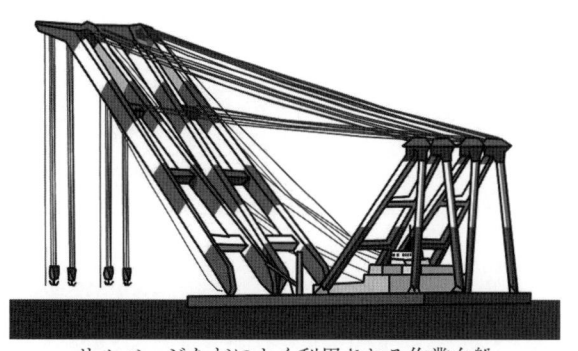

サルベージなどによく利用される作業台船

Unit 11　海難救助捜索 1

Step Check Box □ □ □

Step 1　下記の英文を読み，次の3つの活動をしなさい。終了後 Step Check Box □をぬりなさい。

(1) 動詞（述語動詞）を○で囲む。動詞の意味を考えて，動詞のあとに続く内容を推測する。
(2) 主語（主部）に下線を引く。
(3) スラッシュを入れる。スラッシュを入れる場所は主に下記の3つ。
- コンマ(,)やピリオド(.)でスラッシュ(/)を入れる。音読するときは，スラッシュで息つぎする。
- 接続詞の前にスラッシュを入れる。［接続詞＋主語＋動詞］は文としてまとまった意味を表す。
  ここでいう接続詞とは when, if, until, before, after など (and, but, or, so を含まない)。
- 原則として前置詞の前にスラッシュを入れる。（前置詞＋名詞）は，最小の意味のまとまりを表す。

1) Sector Search
2) Most effective when the position of the search object is accurately known and the search area is small.
3) Used to search a circular area centred on a datum point.
4) Due to the small area involved, this procedure must not be used simultaneously by multiple aircraft at similar altitudes or by multiple vessels.
5) A suitable marker (for example, a smoke float or a radio beacon) may be dropped at the datum position and used as a reference or navigational aid marking the centre of the pattern.
6) For vessels, the search pattern radius is usually between 2 NM and 5 NM, and each turn is 120°, normally turned to starboard.

出典：2級海技士（航海）平成25年10月定期試験問題
IAMSAR Manual（国際航空海上捜索救助マニュアル）Vol. 3 Section 3-22 からの抜粋

Step 2　網掛けは主語，ゴシック体は動詞（述語動詞），＊語注・文法的留意事項，①②③は訳順番号

(1) 単語や熟語の説明を読みながら，部分訳を書きなさい。わからない単語の意味は調べなさい。
(2) 部分訳が終わったら，①②③の訳順番号を参考に，全文和訳を書きなさい。
　全文和訳が終わったら，Step Check Box □をぬりなさい。

1) Sector Search
　　＊sector search：レーダーによる扇形捜索（有効範囲内の捜索）
　　＊レーダーによる扇形捜索とは何か，そのときの注意点は何か，という専門知識を思い出すと英文がわかりやすくなる。

タイトル和訳　_____

2) (Sector search is) Most effective/ [when the position of the search object is accurately known]/ and [the search area is small.]/
（⑤　①　②　③　④）

- ＊sector search is：ここでは省略されている。
- ＊effective：〔形容詞〕効果的な
- ＊search object：捜索対象物
- ＊accurately：〔副詞〕正確に
- ＊search：〔名詞〕捜索，〔他動詞〕〜を探す
- ＊most effective：最上級で書かれているので「最も有効である」と訳す。
- ＊object：〔名詞〕対象物（人），物体，目標
- ＊accurate：〔形容詞〕
- ＊接続詞 and は，何と何をつなぐか？→ ①②と④をつなぐ。

全文和訳　レーダーによる扇形捜索は _____

3) (Sector search is) Used/ to search a circular area/ centred on a datum point./
（④　③　②　①）

- ＊is used：〔受動態〕〜に使われる
- ＊to search：不定詞には「〜することは」「〜するための」「〜するために」の3つの訳し方がある。
　　ここでは不定詞は動詞 is used にかかるので，「〜するために」と訳す。
- ＊a circular area：円形区域
- ＊2)の英文同様に，主語Sと動詞V (Sector search is) が省略されている。
- ＊circular：〔形容詞〕円形の，一周する
- ＊circle：〔名詞〕

48

Unit 11 海難救助捜索 1

* datum：〔名詞〕基準面，基準点，位置評定基準。複数形は data である。
* centred on a datum point：推定基点を中心とする
* centred は過去分詞（受身の意味を持つ）で，形容詞のように名詞 area にかかる。center の意味は「～を中心に置く」。
* centre：イギリス英語のスペル。center はアメリカ英語のスペル。
  海事関連英語では，イギリス英語のスペルが用いられることがとても多い。

全文和訳　レーダーによる扇形捜索は _____

4) Due to the small area involved,/ ① this procedure **must not be used** ② simultaneously/

* due to ～：〔前置詞〕～の理由で ＝ because of
* involved：〔形容詞〕入り組んだ，複雑な　　　　　　* involved は直前の名詞 area を修飾している。
* procedure：〔名詞〕行動（捜索）
* simultaneously：〔副詞〕いっせいに，同時に　（ここでは，下記の③あるいは⑤との同時の捜索を指す）

①②部分訳 _____

③　　　　　　　　　　　④　　　⑤
by multiple aircraft at similar altitudes/ or by multiple vessels./

* multiple：〔形容詞〕複数の　　　* aircraft：（総称的に）航空機　　　* vessel：（大型の）船
* similar：〔形容詞〕似たような　　* similarity：〔名詞〕類似，相似
* altitude：〔名詞〕高度
* 接続詞 or は，何と何をつなぐか？→ ③あるいは⑤

全文和訳 _____

5) ④ A suitable marker (① for example, ② a smoke float or ③ a radio beacon) ⑥ **may be dropped**/ ⑤ at the datum position/

* suitable marker：適切なマーカー
* 英語では，主語や動詞がまず大きなトピックを述べ，次に詳細な情報が追加されて，具体的になっていく。
* a smoke float：発煙浮信号　　　　　　　　　　* suitable：〔形容詞〕ふさわしい
* a radio beacon：無線ビーコン　　　　　　　　* at the datum position：推定基点上に
* may be dropped：may は助動詞，「～しなければいけない，～されるものとする」という意味。

①②③④部分訳 _____ は，

⑦　　　　　⑪　　　　　　　　⑧　　　　　　　　　　⑩　　　⑨
and (a suitable marker **may be**) used/ as a reference or navigational aid/ marking the centre of the pattern./

* 接続詞 and は，何と何をつなぐか？→ may be dropped と may be used をつなぐ。
* as a reference or navigational aid：表示あるいは航路標識として
* reference：〔名詞〕表示，指示，参照
* marking the centre of the pattern：捜索パタン（扇形捜索）の中心を示している
* marking：〔現在分詞〕直前の名詞 aid にかかる。　　* mark：〔動詞〕（旋回軸などを）示す，～の印となる

全文和訳 _____

6) ① For vessels,/ ② the search pattern radius ④ **is** usually/ ③ between 2 NM and 5 NM,/ and ⑤ each turn ⑧ **is** ⑦ 120°, normally ⑧ **turned** ⑥ to starboard./

* the search pattern radius：捜索パタンの半径の範囲

* radius：〔名詞〕半径（r, rad と略す）　　* diameter：〔名詞〕直径
* is：〔自動詞〕あります　　　　　　　　　* between 2 NM and 5 NM：2 海里と 5 海里の間に
* each turn：それぞれの旋回　　　　　　　* is 120° turned：120 度の方向に旋回する
* normally：〔副詞〕普通，正常に　　　　 * normal：〔形容詞〕　　* starboard：〔名詞〕右舷

①②③④部分訳 _____

全文和訳

Step 3　　網掛け は主語，**ゴシック体**は動詞（述語動詞）

(1) 音声を聞きながら，音読する。音読が終わったら，Step Check Box □をぬりなさい。
(2) 左から順に，スラッシュ毎の意味を考える。

Sector Search

(a) Most effective/ when the position of the search object **is** accurately **known**/ and the search area **is** small./

(b) **Used**/ to search a circular area/ centred on a datum point./

(c) Due to the small area involved,/ this procedure **must not be used** simultaneously/ by multiple aircraft at similar altitudes/ or by multiple vessels./

(d) A suitable marker (for example, a smoke float or a radio beacon) **may be dropped**/ at the datum position/ and **used**/ as a reference or navigational aid/ marking the centre of the pattern./

(e) For vessels,/ the search pattern radius **is** usually/ between 2 NM and 5 NM,/ and each turn **is** 120°, normally **turned** to starboard./

【専門英語単語リスト】書いておぼえよう

発音したり，スペル練習したり，意味を調べたりしたらチェック欄の○をぬりなさい。
辞書を引いて単語の意味が多くある場合，英語の文脈から判断し，適切な意味を選びなさい。

英文	単語	品詞	意味を調べて書きなさい	発音しながらスペル練習しなさい	チェック
1)	search	名/動			○○○
2)	effective	形容詞			○○○
2)	object	名詞			○○○
2)	area	名詞			○○○
3)	circular	形容詞			○○○
3)	datum / data（複数形）	名詞			○○○
4)	due to	熟語			○○○
4)	involve	他動詞			○○○
4)	procedure	名詞			○○○
4)	altitude	名詞			○○○
5)	navigational	形容詞			○○○
5)	aid	名詞			○○○
5)	mark	他動詞			○○○
6)	radius / diameter	名詞			○○○
6)	starboard	名詞			○○○

理解を深めるキーワード「SAR」

　海洋上における人命救助は，捜索範囲が広範囲であるにもかかわらず短時間での捜索を要求されるため，難易度が高く，また，外洋上では国をまたぐ場合がある。そこで，効率よく救助を進めるための，救助隊の役割や通信手段などについての国際的な取り決めが必要であり，現在ではSAR（Search and Rescuer）条約が発効している。その条約を批准（ルールを守ると宣言すること）する国は，各国領海の周辺国とともにSAR協定と呼ばれる，海洋上における人命救助に関する協定を結び，迅速な遭難船の発見（船位通報），そして救助を行うことができるように整備されている。その際の手順，ルールマニュアルが「IAMSAR」（国際航空海上捜索救助マニュアル）である。本マニュアルには捜索する手順について細かに示されており，「拡大方形捜索（図1）」，2隻の船舶・航空機を併用することで捜索効率が向上する「扇形捜索（図2）」，何隻かで合同で行う「平行スイープ捜索，平行トラック捜索（図3）」，航空機と合同で行う「合同クリープ線捜索（図4）」と，大きく分類して4つの捜索方法が掲載されている。

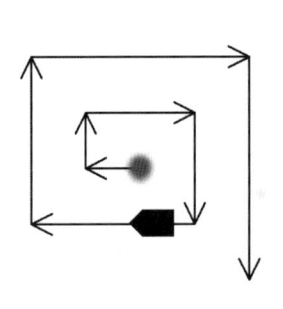

図1　拡大方形捜索

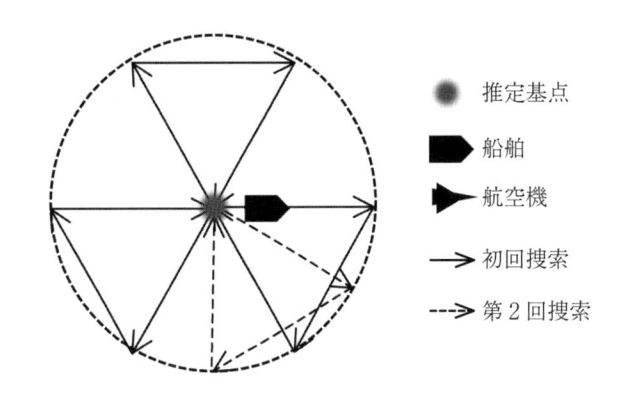

図2　扇形捜索

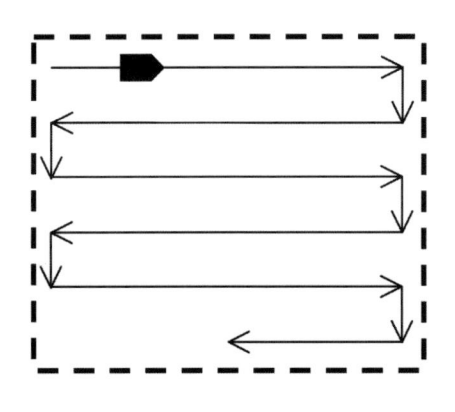

図3　平行スイープ，トラック捜索

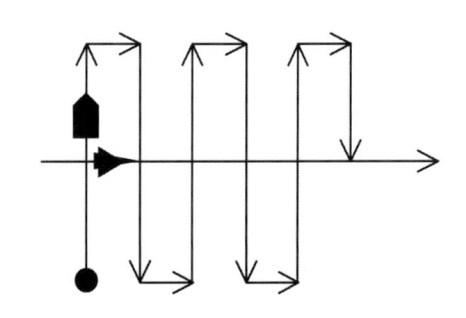

図4　合同クリープ線捜索

Unit 12　海難救助捜索 2

Step Check Box ☐ ☐ ☐

Step 1　下記の英文を読み，次の3つの活動をしなさい。終了後 Step Check Box ☐ をぬりなさい。

(1) 動詞（述語動詞）を○で囲む。動詞の意味を考えて，動詞のあとに続く内容を推測する。
(2) 主語（主部）に下線を引く。
(3) スラッシュを入れる。スラッシュを入れる場所は主に下記の3つ。
- コンマ(,)やピリオド(.)でスラッシュ(/)を入れる。音読するときは，スラッシュで息つぎする。
- 接続詞の前にスラッシュを入れる。〔接続詞＋主語＋動詞〕は文としてまとまった意味を表す。
  ここでいう接続詞とは when, if, until, before, after など（and, but, or, so を含まない）。
- 原則として前置詞の前にスラッシュを入れる。（前置詞＋名詞）は，最小の意味のまとまりを表す。

1) Recovery of survivors by assisting vessels
2) Seafarers should consider how to recover survivors into their own vessels under various environmental conditions.
3) Survivors in the water should be lifted in a horizontal or near-horizontal position if possible (for example, in two strops; one under the arms, the other under the knees) to minimize the risk of shock induced by sudden transfer from the water and possible hypothermia.
4) Assisting vessels should also be prepared to receive survivors from helicopters: see pages 2-50.
5) When the risks involved in recovery operations outweigh the risks of leaving the survivors in life saving appliances, consider the following actions:
6) Using the ship to provide a lee for the survivors;
7) Deploying life saving appliances from the assisting vessel;
8) Maintaining visual and communications contact with the survivors;
9) Updating the co-ordinating authority;
10) Transferring essential survival and medical supplies.

出典：2級海技士（航海）平成26年2月定期試験問題
IAMSAR Manual（国際航空海上捜索救助マニュアル）Vol. 3 Section2-36 からの抜粋

Step 2　網掛けは主語，**ゴシック体**は動詞（述語動詞），＊語注・文法的留意事項，①②③は訳順番号

(1) 単語や熟語の説明を読みながら，部分訳を書きなさい。わからない単語の意味は調べなさい。
(2) 部分訳が終わったら，①②③の訳順番号を参考に，全文和訳を書きなさい。
　全文和訳が終わったら，Step Check Box ☐ をぬりなさい。

1)　　②　　　　　　①
　Recovery of survivors/ by assisting vessels

* タイトル：タイトルがわかると，書かれている英文全体が推測できる。タイトルの後に具体的で詳細な情報が述べられる。
* recovery：〔名詞〕取り戻すこと，ここでは「収容」を指す。　＊ recover：〔他動詞〕～を取り戻す
* survivor：〔名詞〕生存者　　　　　　　　　　　　　　＊ survive：〔自動詞〕　　　＊ survival：〔形容詞・名詞〕
* assist：〔他動詞〕助ける＝help　　　　　　　　　　　　＊ assistance：〔名詞〕
* by：〔前置詞〕前置詞の後には必ず名詞が置かれるので，動詞 assist を動名詞（Ving）に変化させる。
* assisting vessel：救助船，支援船舶

タイトル和訳　_____

2)　①　　　　⑤　　　　　　④　　　　　　　　　　　③
　Seafarers **should consider** how to recover survivors/ into their own vessels/
　　　　　　　　　　　　　　　　　②
　under various environmental conditions./

* should consider の後の英文には，その consider する内容・中身に関することが書かれている。
　英文は，左から右へと情報がだんだん詳細に具体的に，追加されていく。
* how to 動詞の原形：～の仕方，いかに～するべきか　　＊（疑問詞＋不定詞）の訳し方に注意。
* into：〔前置詞〕～のなかへ　　　　　　　　　　　　　＊ own：〔形容詞〕～自身の，〔動詞〕所有する

52

* under：〔前置詞〕～の下に，（影響）下で
* various：〔形容詞〕さまざまな
* environmental：〔形容詞〕環境の，周囲の
* conditional：〔形容詞〕条件付きの

* under には何かに覆われているというイメージがある。
* variety：〔名詞〕多様性　　　* vary：〔動詞〕
* environment：〔名詞〕自然環境
* condition：〔名詞〕

全文和訳

3) Survivors in the water **should be lifted**/ in a horizontal (position) or (a) near-horizontal position/
　　①　　　　　　　　　⑬　　　　　　　　　　⑪　　　　　　　　　　⑫

* in the water：水中の
* should be lifted：〔受動態〕助動詞の後には必ず動詞の原形が付く。　　* lift：〔他動詞〕（重いものを）持ち上げる
* horizontal：〔形容詞〕水平線の　　* in a horizontal position：水平の（横になった）姿勢で
* horizon：〔名詞〕　　* near-：〔接頭語〕ほぼ，ほとんど

①⑪⑫⑬部分訳 _____

[if possible for example,/ in two strops;/ one under the arms,/ the other under the knees/]
　②　　　　　　　　③　　　　　　④　　　　　　⑤　　　　　　　⑥

* if (it is) possible：〔熟語〕もし可能ならば。[if possible 以下]は，⑬ should be lifted につながる。
* in two strops：2つのストロップ（ループ）で
* strop：〔名詞〕ストロップ（滑車のフックを引っかけるための帯索（環索））
* two strops：⑤と⑥で，具体的に書いてある。「ひとつは…にかけて，もうひとつは…にかけて」と訳す。
* arm：〔名詞〕腕　　* knee：〔名詞〕ひざ

②③④⑤⑥部分訳 _____

to minimize the risk of shock/ induced by sudden transfer from the water and possible hypothermia./
　　⑩　　　　　　　　　　　　⑨　　　　　　　　　　⑦　　　　　　　　　　　　⑧

* minimize：〔他動詞〕～を最小限にする　　* to minimize：〔不定詞〕～するために
* shock：〔名詞〕ショック，心停止
* induce：〔他動詞〕～を引き起こす，誘発する
* induced：〔過去分詞〕過去分詞の前に名詞 shock が付いているので，induced by sudden transfer は，shock を修飾する形容詞と同じ働きをする。
* sudden：〔形容詞〕突然の，不意の　　* suddenly：〔副詞〕
* transfer：〔名詞〕移動，乗換，〔他動詞〕～を移動させる，伝達する
* possible：〔形容詞〕（可能性として）ありうる　　* possibility：〔名詞〕可能性
* hypothermia：〔名詞〕低体温症　　* hypo-：〔接頭語〕低い，下位の，過小の

全文和訳

4) Assisting vessels **should** also **be prepared**/ to receive survivors from helicopters:/ see pages 2-50./
　　　　①　　　　　　　④　　　　　　　　　③　　　　　　　　　②　　　　　　⑤

* assisting：〔現在分詞〕救助している，形容詞のように名詞 vessel にかかる。
* to receive：不定詞は動詞の意味を持ち，働きが3つある。「～することは」（＝名詞と同じ働き），「～するための」（＝形容詞と同じ働き），「～するために」（＝副詞と同じ働き）。
* see：命令文で，「（2～50ページを）参照しなさい」となる。

全文和訳

5) [When the risks involved in recovery operations **outweigh** the risks of leaving the survivors/
　　　　②　　　　　　①　　　　　　　　　　　　　⑥　　　　　　⑤　　　　　　　④
in life saving appliances],/ **consider** the following actions:/
　　　③　　　　　　　　　⑧　　　⑦

* when：〔接続詞〕～のときに，when の後に主語 S と動詞 V が付く。
* involved in：〔過去分詞〕～に含まれる，名詞 the risks を修飾する。
* the risks involved in recovery operations：収容作業に含まれる危険

* operation：〔名詞〕操作，作業
* survivor：〔名詞〕生存者
* leaving：〔動名詞〕残すこと
* consider：〔他動詞〕～を配慮する，考慮する （ここでは命令文なので「～しなさい」と訳す）
* action：〔名詞〕行動
* outweigh：〔動詞〕（重みが）～以上である，（重要さが）～にまさる
* in life saving appliances：人命救助装置のなかに
* 前置詞 of の後に動詞を置くときは，ing形（動名詞）にする。
* the following actions：これは 6) ～ 10) の action を指す。

①②部分訳＿＿＿＿＿＿＿＿＿＿＿＿＿＿＿＿＿＿　③④⑤部分訳＿＿＿＿＿＿＿＿＿＿＿＿＿＿＿＿

⑦⑧部分訳＿＿＿＿＿＿＿＿＿＿＿＿＿＿＿＿＿＿

全文和訳 ＿＿

6) Using the ship/ to provide a lee/ for the survivors;/
　　　③　　　　　②　　　　　　①

* using：〔動名詞〕動詞の意味を持っているが，働きは名詞と同じ。「～を使用すること」と訳す。
* to provide：用意するために
* 不定詞は動詞の意味を持ち，働きが3つある。「～することは」(= 名詞と同じ働き)，「～するための」(= 形容詞と同じ働き)，「～するために」(= 副詞と同じ働き)。
* provide：〔他動詞〕= supply
* lee：〔名詞〕風下 ↔ weather side = windward：風上

全文和訳 ＿＿

7) Deploying life saving appliances/ from the assisting vessel;/
　　　③　　　　　　　②　　　　　　　　　　①

* deploy：〔他動詞〕～を配備する，展開させる
* life saving appliances：人命救助装置
* appliance：〔名詞〕装置，適用，電気器具
* deploying：〔動名詞〕～を配備すること
* apply：〔他動詞〕適合する，出願する，志望する

全文和訳 ＿＿

8) Maintaining visual (contact) and communications contact/ with the survivors;/
　　　④　　　　②　　　　　　　　　　③　　　　　　　　　　　①

* maintain：〔他動詞〕持続する，維持する
* visual (contact)：視覚による観察，ここでは visual contact with と続く。
* communication contact with：～との直接連絡
* maintaining：〔動名詞〕持続すること
* visible：〔形容詞〕目に見える

全文和訳 ＿＿

9) Updating the co-ordinating authority;/
　　　②　　　　　　①

* update：〔他動詞〕～の最新情報を提供する
* co-ordinating authority：調整機関
* co-ordinate：〔他動詞〕（部分，働きなどを）調整する

全文和訳 ＿＿

10) Transferring essential survival (supplies) and medical supplies./
　　　　③　　　　　　①　　　　　　　　　　②

* transfer：〔他動詞〕輸送する，積み換える = move, convey（ここでは積み換えするものが2つ記述されている）
* essential：〔形容詞〕最も重要な，必須の = necessary
* medical：〔形容詞〕医療の
* supply：〔名詞〕供給品，補給，蓄え ↔ demand：〔名詞〕需要
* survival：〔形容詞〕緊急・非常時用の
* medicine：〔名詞〕

全文和訳 ＿＿

Step 3 網掛け は主語，**ゴシック体**は動詞（述語動詞）

(1) 音声を聞きながら，音読する。音読が終わったら，Step Check Box □をぬりなさい。
(2) 左から順に，スラッシュ毎の意味を考える。

Recovery of survivors/ by assisting vessels

- Seafarers **should consider** how to recover survivors/ into their own vessels/ under various environmental conditions./
- Survivors in the water **should be lifted**/ in a horizontal or near-horizontal position/ if possible for example,/ in two strops;/ one under the arms,/ the other under the knees/ to minimize the risk of shock/ induced by sudden transfer from the water and possible hypothermia./
- Assisting vessels **should** also **be prepared**/ to receive survivors from helicopters;/ see pages 2-50./
- When the risks involved in recovery operations **outweigh** the risks of leaving the survivors/ in life saving appliances,/ **consider** the following actions:/
  a) Using the ship/ to provide a lee/ for the survivors;/
  b) Deploying life saving appliances/ from the assisting vessel;/
  c) Maintaining visual and communications contact/ with the survivors;/
  d) Updating the co-ordinating authority;/
  e) Transferring essential survival and medical supplies./

【専門英語単語リスト】書いておぼえよう

発音したり，スペル練習したり，意味を調べたりしたらチェック欄の○をぬりなさい。
辞書を引いて単語の意味が多くある場合，英語の文脈から判断し，適切な意味を選びなさい。

英文	単語	品詞	意味を調べて書きなさい	発音しながらスペル練習しなさい	チェック
1)	recover / recovery	動 / 名			○○○
1)	assist / assistance	動 / 名			○○○
2)	consider	他動詞			○○○
3)	lift	他動詞			○○○
3)	horizon / horizontal	名 / 形			○○○
3)	strop	名詞			○○○
3)	minimize	他動詞			○○○
3)	induce	他動詞			○○○
4)	survivor / survival	名詞			○○○
5)	involve	他動詞			○○○
6)	provide	他動詞			○○○
7)	deploy	他動詞			○○○
8)	contact	名 / 動			○○○
9)	update	他動詞			○○○
10)	supply	動 / 名			○○○

Unit 13　救命いかだ

Step Check Box ☐ ☐ ☐

Step 1　下記の英文を読み，次の3つの活動をしなさい。終了後 Step Check Box ☐ をぬりなさい。

(1) 動詞（述語動詞）を○で囲む。動詞の意味を考えて，動詞のあとに続く内容を推測する。
(2) 主語（主部）に下線を引く。
(3) スラッシュを入れる。スラッシュを入れる場所は主に下記の3つ。
- コンマ(,)やピリオド(.)でスラッシュ(/)を入れる。音読するときは，スラッシュで息つぎする。
- 接続詞の前にスラッシュを入れる。［接続詞＋主語＋動詞］は文としてまとまった意味を表す。
  ここでいう接続詞とは when, if, until, before, after など（and, but, or, so を含まない）。
- 原則として前置詞の前にスラッシュを入れる。（前置詞＋名詞）は，最小の意味のまとまりを表す。

1) Stability of inflatable liferafts
2) Every inflatable liferaft shall be so constructed that, when fully inflated and floating with the canopy uppermost, it is stable in a seaway.
3) The stability of the liferaft when in the inverted position shall be such that it can be righted in a seaway and in calm water by one person.
4) The stability of the liferaft when loaded with its full complement of persons and equipment shall be such that it can be towed at speeds of up to 3 knots in calm water.
5) The liferaft shall be fitted with water pockets complying with the following requirements:
5-1) the water pockets shall be of a highly visible colour;
5-2) the pockets shall be positioned symmetrically round the circumference of the liferaft.
6) Means shall be provided to enable air to readily escape from underneath the liferaft.

出典：2級海技士（航海）平成26年4月定期試験問題
International Life-Saving Appliance (LSA) Code（国際救命設備コード）からの抜粋

Step 2　網掛け は主語，**ゴシック体**は動詞（述語動詞），＊語注・文法的留意事項，①②③は訳順番号

(1) 単語や熟語の説明を読みながら，部分訳を書きなさい。わからない単語の意味は調べなさい。
(2) 部分訳が終わったら，①②③の訳順番号を参考に，全文和訳を書きなさい。
   全文和訳が終わったら，Step Check Box ☐ をぬりなさい。

1) Stability of inflatable liferafts

　＊救命ボートを膨らませ安定した状態にしておくための方法が予備知識として必要
　＊ stability：〔名詞〕安定性　　　　　　　　　　＊ stable：〔形容詞〕安定した
　＊ inflatable：〔形容詞〕膨らませることができる　＊ inflated：〔形容詞〕（空気・気体で）膨らんだ
　＊ liferaft：〔名詞〕救命いかだ＝ life raft　　　　＊ inflatable liferafts：膨張式救命いかだ

タイトル和訳
_____

2) Every inflatable liferaft① **shall be** so **constructed**⑦/ [that, when (it is) fully **inflated**② and (it is) **floating**④/ with the canopy uppermost③,/ it **is**⑥ stable⑤ in a seaway.]/

　＊ shall be so constructed [that … it is stable in a seaway]：that 以下になるようにつくられるべきだ
　＊ when (it is) fully inflated：(liferaft が)十分に膨らんでいて　(it is が省略)
　＊ (it is) floating with the canopy uppermost：最も高い所に (uppermost) 天蓋 (canopy) を付けて浮いている
　＊ [that (when … uppermost) it is stable in a seawa]：海中で (救命ボートが) 安定するように
　＊ in a seaway：海水のなかで　　　　　　　　　＊ construct：〔他動詞〕～をつくる，建設する

全文和訳
_____

3) The stability of the liferaft/ [when (it **is**) in the inverted position]/ **shall be** such [that **it can be righted**/

- * invert：〔他動詞〕ひっくり返す
- * in the inverted position：ひっくり返った状態で
- * inverted は過去分詞で, position にかかる。
- * such that 主語 S ＋動詞 V：such は代名詞で「…のようなもの」という意味。
  such that は熟語で,「that 以下のようなもの」と訳す。
- * right：〔他動詞〕転覆したものをまっすぐにする, 船・ボートを水平に立て直す
- * can be righted：〔受動態〕まっすぐになる（される）　　* righted と ed が付いているので, right は動詞である
- * 全体の大意は「救命ボートの安定は, ②の状態のとき, that 以下のようなものにすべきである」となる。

①②部分訳 _____

in a seaway and in calm water/ by one person.] /

- * in a seaway and in calm water：海中で波が穏やかなときに　　* calm：〔形容詞〕おだやかな
- * by one person：ひとりで

全文和訳 _____

4) The stability of the liferaft/ [when (it **is**) **loaded with** its full complement of persons and equipment]/ **shall be** such [that **it can be towed**/ at speeds of up to 3 knots/ in calm water.]/

- * be loaded with：〔受動態〕with 以下が載せられている　　* load：〔他動詞〕（荷などを）載せる
- * complement：〔名詞〕船の乗組定員　　* equipment：〔名詞〕備品
- * when 以下①②③の大意は「乗組員や備品などで乗組定員いっぱいが救命いかだにのっているときは」となる。
- * shall be such that 主語 S ＋動詞 V：such は代名詞で「…のようなもの」という意味。
  such that は熟語で,「that 以下のようなもの」と訳す。
- * can be towed：〔受動態〕曳航される　　* at speeds of ～：～の速さで
- * up to ～：〔熟語〕～（の深さや高さ）までの, ～にいたるまで

全文和訳 _____

5) The liferaft **shall be fitted with** water pockets/ complying with the following requirements:/

- * shall be fitted with：with 以下が備え付けられるべきである
- * water pocket：安定水のう（水が入る袋）
- * comply with：(基準などを)満たす　　* complying with：現在分詞以下が water pockets にかかる。
- * the following requirements：下記の必要条件, 要件

全文和訳 _____

5-1) the water pockets **shall be**/ of a highly visible colour;/

- * highly：〔副詞〕かなり　　* visible：〔形容詞〕目に見える
- * colour：イギリス英語のスペル。アメリカ英語では color。
- * of colour：〔熟語〕カラフルな, 色彩に富んだ＝ colourful
- * 海技士試験では, イギリス英語のスペルが頻繁に用いられる。

全文和訳 _____

5-2) the pockets **shall be positioned** symmetrically/ round the circumference of the liferaft./
　　　①　　　　　　　⑤　　　　　　　　④　　　　③　　　　　　　　　②

* shall be positioned：〔受動態〕置かれるべきである
* symmetrically：〔副詞〕対称に
* round：〔前置詞〕をぐるりと取り巻いて，を囲んで
* circumference：〔名詞〕周囲

全文和訳

6) Means **shall be provided**/ to enable air to readily escape/ from underneath the liferaft./
　　①　　　　　④　　　　　　　　　　　③　　　　　　　　　　　②

* means：〔名詞〕（複数形で）手段，方法
* shall be provided：（手段が）用意されるべきである
* readily：〔副詞〕容易に
* escape from：〔熟語〕～から逃げる
* enable A to B：AがBできるようにする
* to enable air to readily escape：空気が容易に抜け出るように
* underneath：〔前置詞〕の真下に

全文和訳

Step 3   網掛けは主語，**ゴシック体**は動詞（述語動詞）

(1) 音声を聞きながら，音読する。音読が終わったら，Step Check Box □をぬりなさい。
(2) 左から順に，スラッシュ毎の意味を考える。

5　Stability of inflatable liferafts

5.1　Every inflatable liferaft **shall be** so **constructed**/ that, when fully **inflated** and **floating**/ with the canopy uppermost,/ it **is** stable in a seaway./

5.2　The stability of the liferaft/ when in the inverted position/ **shall be** such that it **can be righted**/ in a seaway and in calm water/ by one person./

5.3　The stability of the liferaft/ when **loaded with** its full complement of persons and equipment/ **shall be** such that it **can be towed**/ at speeds of up to 3 knots/ in calm water./

5.4　The liferaft **shall be fitted with** water pockets/ complying with the following requirements:/
　　1　the water pockets **shall be**/ of a highly visible colour;/
　　2　the pockets **shall be positioned** symmetrically/ round the circumference of the liferaft./

Means **shall be provided**/ to enable air to readily escape/ from underneath the liferaft./

【文法のポイント】
接続詞の前，原則として前置詞の前で，スラッシュを入れる。スラッシュは，意味のまとまりを示す。スラッシュ毎に，左から右へ区切りながら訳せば，英文の内容が理解できる。和訳をする場合は，下記のポイントを参考に，日本語らしい語順に直す。
1. ［接続詞＋主語S＋動詞V…］は，動詞Vにかかる。
2. （前置詞＋名詞（もの・コト））は，直前（直後）の名詞や動詞にかかる。
3. ［関係代名詞＋主語S＋動詞V…］は，直前の名詞にかかる。
4. 名詞＋過去分詞（Ved）の語順の場合，過去分詞（Ved）は，直前の名詞にかかる。
5. 名詞＋現在分詞（Ving）の語順の場合，現在分詞（Ving）は，直前の名詞にかかる。

【専門英語単語リスト】書いておぼえよう

発音したり，スペル練習したり，意味を調べたりしたらチェック欄の○をぬりなさい。
辞書を引いて単語の意味が多くある場合，英語の文脈から判断し，適切な意味を選びなさい。

英文	単語	品詞	意味を調べて書きなさい	発音しながらスペル練習しなさい	チェック
1)	stability / stable	名/形			○○○
1)	inflatable / inflate	形/動			○○○
1)	liferaft	名詞			○○○
2)	construct	他動詞			○○○
2)	float	自動詞			○○○
3)	invert	他動詞			○○○
3)	right	動/名			○○○
4)	load	他動詞			○○○
5)	comply with	自動詞			○○○
5)	requirement	名詞			○○○
5-1)	visible / visibility	形/名			○○○
5-2)	circumference	名詞			○○○
6)	means	名詞			○○○
6)	enable / able	動/形			○○○
6)	escape from	自動詞			○○○

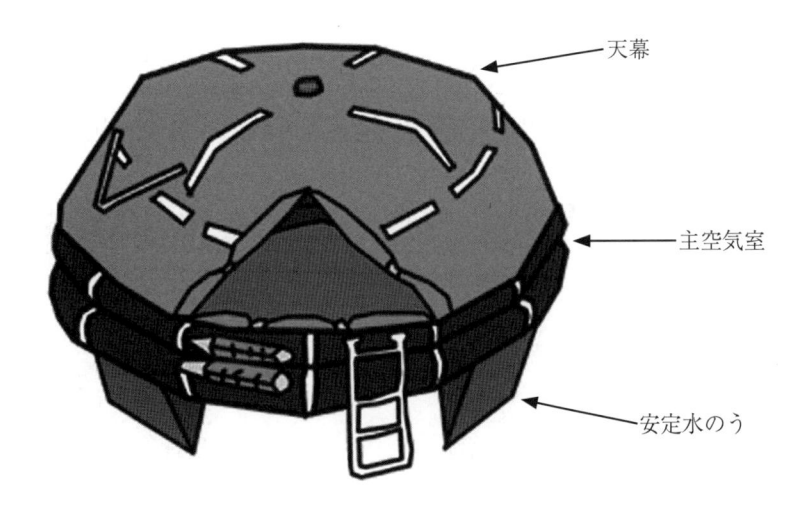

理解を深めるキーワード「救命いかだ」

- 天幕
- 主空気室
- 安定水のう

Unit 14 　積荷に対する責任　　　　　Step Check Box □ □ □

Step 1　下記の英文を読み，次の3つの活動をしなさい。終了後 Step Check Box □をぬりなさい。

(1) 動詞（述語動詞）を○で囲む。動詞の意味を考えて，動詞のあとに続く内容を推測する。
(2) 主語（主部）に下線を引く。
(3) スラッシュを入れる。スラッシュを入れる場所は主に下記の3つ。
- コンマ(,)やピリオド(.)でスラッシュ(/)を入れる。音読するときは，スラッシュで息つぎする。
- 接続詞の前にスラッシュを入れる。［接続詞＋主語＋動詞］は文としてまとまった意味を表す。
- 原則として前置詞の前にスラッシュを入れる。（前置詞＋名詞）は，最小の意味のまとまりを表す。

1) Cargo and cargo units carried on or under deck shall be so loaded, stowed and secured as to prevent as far as is practicable, throughout the voyage, damage or hazard to the ship and the persons on board, and loss of cargo overboard.

2) Cargo carried in a cargo unit shall be so packed and secured within the unit as to prevent, throughout the voyage, damage or hazard to the ship and the persons on board.

3) Appropriate precautions shall be taken during loading and transport of heavy cargoes or cargoes with abnormal physical dimensions to ensure that no structural damage to the ship occurs and to maintain adequate stability throughout the voyage.

4) Containers shall not be loaded to more than the maximum gross weight indicated on the Safety Approval Plate under the International Convention for Safe Containers.

出典：2級海技士（航海）平成21年2月定期試験問題
SOLAS条約 Chapter 4, Regulation 5 からの抜粋

Step 2　網掛けは主語，**ゴシック体**は動詞（述語動詞），＊語注・文法的留意事項，①②③は訳順番号

(1) 単語や熟語の説明を読みながら，部分訳を書きなさい。わからない単語の意味は調べなさい。
(2) 部分訳が終わったら，①②③の訳順番号を参考に，全文和訳を書きなさい。
　全文和訳が終わったら，Step Check Box □をぬりなさい。

1) ②Cargo and cargo units/ ①carried on (deck) or under deck **shall be** so ⑨**loaded, stowed and secured**

- ＊ cargo：〔名詞〕荷物，積荷
- ＊ cargo unit：貨物ユニット，コンテナ
- ＊「主語と動詞」の意味・内容がわかれば，どのような英文があとに続くのかを推測するのが容易になる。
- ＊ cargo and cargo units carried on：過去分詞 carried が，直前の名詞 cargo and cargo units を修飾する。carried は過去分詞なので「運ばれる」と受身のように訳す。
- ＊ load：〔他動詞〕荷を載せる
- ＊ stow：〔他動詞〕（容器に）〜を積み込む
- ＊ secure：〔他動詞〕〜を固定する
- ＊ shall be Ved：〜されるべきである
- ＊ shall be so loaded：下記の⑧へ，so as to V として続く。

①②部分訳＿＿＿＿＿＿＿＿＿＿＿＿＿＿＿＿＿＿＿＿＿＿＿＿＿＿＿＿＿＿＿＿＿＿＿＿＿

⑧as to prevent/ ③as far as (it) **is** practicable,/ ④throughout the voyage,/

- ＊ so 〜 as to prevent：〜ように，〜するために＝in order to V
- ＊ as to prevent：防止するのは⑤⑥と⑦
- ＊ as far as it is practicable：実際的に可能な限り
- ＊ as far as：（接続詞的な使い方）〜の限り
- ＊ practicable：〔形容詞〕実用的な，実行可能な
- ＊ throughout the voyage：航海の間

⑥damage or hazard to the ship and the persons on board,/ ⑦and loss of cargo overboard.

- ＊ damage or hazard：(⑤〜に対して加えられた) 損害と危険
- ＊ the persons on board：船上にいる人＝乗船者
- ＊ loss of cargo overboard：船外への積荷の損失 →「船外に積荷が落ち損失すること」と解釈できる。

⑤⑥⑦部分訳 _____

⑧⑨部分訳 _____

* この英文の主語は①②，動詞は⑨である。
* 動詞⑨は「貨物を積載し，詰め込み，固定すべきだ」という意味。その意味から，後に続く英文の内容を推測することができる。⑤⑥⑦⑧には，なぜ⑨のようにするのかという情報が，具体的に書かれている。

全文和訳
_____

2) ② ① ⑨ ⑧ ⑦
Cargo carried in a cargo unit **shall be** so **packed** and **secured**/ within the unit/ as to prevent,/

* cargo carried in a cargo unit：過去分詞 carried（受身の意味）が直前の名詞 cargo を修飾する。
* pack：〔他動詞〕～を詰める　　　　　　　　* within the unit：コンテナ内部で
* shall be so packed and secured as to prevent：so ～ as to prevent は「守る（防止する）ために」という意味。
* prevent：「防止する，守る」という意味なので，そのあとには「何を防止するのか」という情報④⑤⑥が続く。

①②部分訳 _____　⑦⑧⑨部分訳 _____

③ ⑥ ④ ⑤
throughout the voyage,/ damage or hazard to the ship and the persons on board./

* throughout：〔前置詞〕～を通して（はじめから終わりまでのイメージ）

④⑤⑥部分訳 _____

全文和訳
_____

3) ⑩ ⑪
Appropriate precautions **shall be taken**/

* appropriate precautions：適切な予防措置　　　　* shall be taken：〔受動態〕取られるべきである

④ ① ③ ②
during loading and transport/ of heavy cargoes or cargoes with abnormal physical dimensions/

* during：〔前置詞〕（特定期間）の間に，～の間中
* during loading and transport：積荷をしたり輸送したりしている間に
  「何を」積荷し輸送するのかは，そのあとの①②③に記述してある。
* with abnormal physical dimensions：物理的に異常なほどの大きさがある。この語句は cargoes にかかる。

①②③④部分訳 _____

⑦ ⑤ ⑥
to ensure [that no structural damage to the ship occurs]/ and

* to ensure：[that 以下を] 確認するために　　　　* structural damage to ～：～への構造的な損害
* 主語が否定形（no structural damage）の文の場合は，動詞を打ち消して訳す。
  ここでは，no ～ occurs だから「～が起こらない」と訳す。
* and：〔接続詞〕～と～　（ここでは and は，⑦⑤⑥と⑨⑧をつなぐ働きをしている）

⑤⑥⑦部分訳 _____するために

⑨ ⑧
to maintain adequate stability/ throughout the voyage./

* to maintain adequate stability：十分な安定性を維持するために　　* throughout：〔前置詞〕～の間に
* to ensure 以下と to maintain 以下は，動詞⑪に続く。

全文和訳

4) <u>Containers</u> **shall not be loaded**/
　　①　　　　　　　⑥

* 主語 S ＋動詞 V の後には，SV の内容を具体的，詳細に述べる内容が続く。
　1) の英文の主語，動詞の意味・内容から，「船荷」に関することがトピックになっていることがわかる。
　4) の英文も，引き続き，船荷に関することが書かれている。
* containers：コンテナ
* shall not be loaded：〔受動態〕（コンテナは）船に載せられるべきではない

to more than the maximum gross weight/ indicated on the Safety Approval Plate/
　　　⑤　　　　　　　　　④　　　　　　　　　　　　　　　　　③

* to more than ～：～より以上までに
* maximum gross weight：最大総重量
* indicated：〔過去分詞〕明示されている
* weight indicated on ～：indicated 以下③が前の名詞 weight を修飾する。
* Safety Approval Plate：安全承認プレート，CSC プレート

③④⑤部分訳 _____

under the International Convention for Safe Containers./
　　　　　　　　　　　　　②

* under：〔前置詞〕～のもとで。何かに覆われた「下にある」というイメージ。
* 前置詞の後に必ず名詞が続き，まとまりのある意味をなす。
* International Convention for Safe Containers（＝ CSC）：安全なコンテナに関する国際条約

全文和訳

Step 3　網掛け は主語，**ゴシック体**は動詞（述語動詞）

(1) 音声を聞きながら，音読する。音読が終わったら，Step Check Box □をぬりなさい。
(2) 左から順に，スラッシュ毎の意味を考える。

Cargo and cargo units/ carried on or under deck **shall be** so **loaded, stowed and secured** as to prevent/ as far as **is** practicable,/ throughout the voyage,/ damage or hazard to the ship and the persons on board,/ and loss of cargo overboard./
Cargo carried in a cargo unit **shall be** so **packed** and **secured**/ within the unit/ as to prevent,/ throughout the voyage,/ damage or hazard to the ship and the persons on board./
Appropriate precautions **shall be taken**/ during loading and transport/ of heavy cargoes or cargoes with abnormal physical dimensions/ to ensure that no structural damage to the ship **occurs**/ and to maintain adequate stability/ throughout the voyage./
Containers **shall not be loaded**/ to more than the maximum gross weight/ indicated on the Safety Approval Plate/ under the International Convention for Safe Containers./

【文法のポイント】

海技士試験の英文の特徴は，まず受動態で書かれている文が多い。ここでは，主語が「積荷」なので，4つの英文は，下記のようにすべて受動態で書かれている。

1) Cargo and cargo units/ carried on or under deck/ **shall be so loaded, stowed and secured**
2) Cargo carried in a cargo unit **shall be so packed and secured**/ within the unit
3) Appropriate precautions **shall be taken**/ during loading and transport
4) Containers **shall not be loaded**

- 英文の特徴として，まずトピックが述べられた後に「具体的で詳細な情報が後から後から付け足される」ということを知っておくと，和訳しやすい。たとえば：
  上記の英文1)では積荷方法の注意が述べられ，次に2)では貨物用コンテナ中での積荷方法の注意点を説明している。
  3)では重量貨物や大型貨物の輸送時の注意点を述べた後，4)ではコンテナの重量に関する注意点を述べている。
- 英文では，第1文にいちばん大事な内容（トピック）を述べ，後から具体的で詳細なことが追加されることが多い。

【専門英語単語リスト】書いておぼえよう

発音したり，スペル練習したり，意味を調べたりしたらチェック欄の○をぬりなさい。
辞書を引いて単語の意味が多くある場合，英語の文脈から判断し，適切な意味を選びなさい。

英文	単語	品詞	意味を調べて書きなさい	発音しながらスペル練習しなさい	チェック
1)	cargo	名詞			○○○
1)	load	他動詞			○○○
1)	stow / stowage	動 / 名			○○○
1)	secure	他動詞			○○○
1)	prevent	他動詞			○○○
1)	damage	名詞			○○○
1)	hazard / hazardous	名 / 形			○○○
1)	on board	熟語			○○○
2)	pack	他動詞			○○○
3)	appropriate	形容詞			○○○
3)	precaution	名詞			○○○
3)	transport	名詞			○○○
3)	occur	自動詞			○○○
3)	maintain	他動詞			○○○
3)	stability	名詞			○○○

Unit 15 海難救助と積荷の避難

Step Check Box ☐ ☐ ☐

Step 1 下記の英文を読み，次の3つの活動をしなさい。終了後 Step Check Box □をぬりなさい。

(1) 動詞（述語動詞）を○で囲む。動詞の意味を考えて，動詞のあとに続く内容を推測する。
(2) 主語（主部）に下線を引く。
(3) スラッシュを入れる。スラッシュを入れる場所は主に下記の3つ。
- コンマ(,)やピリオド(.)でスラッシュ(/)を入れる。音読するときは，スラッシュで息つぎする。
- 接続詞の前にスラッシュを入れる。[接続詞＋主語＋動詞]は文としてまとまった意味を表す。
  ここでいう接続詞とは when, if, until, before, after など（and, but, or, so を含まない）。
- 原則として前置詞の前にスラッシュを入れる。(前置詞＋名詞)は，最小の意味のまとまりを表す。

1) Preparation for Cargo Transfer
2) If a ship becomes disabled or stranded, it may be necessary to transfer all or part of her cargo to another vessel.
3) The operation will be greatly expedited if the ship to be lightened makes the following preparations where possible:
4) – Establish contact with the lightening vessel and make a detailed plan of the proposed operation including the designation of a communications channel.
5) Fixed or portable hand-held radio telephones may be usefully employed.
6) – Lay out mooring lines, heaving lines, messengers, stoppers, fenders, etc.
7) If no large fenders are available, mooring ropes and any other soft material should be stung over the ship's side in strategic positions.
8) – Have the anchors cleared ready for use, if in waters where use may be possible.
9) – Brief the officers and crew on the operation, with particular reference to the safety aspects and complete the appropriate safety check list if possible.

出典：2級海技士（航海）平成21年7月定期試験問題

Step 2 　網掛け は主語，**ゴシック体**は動詞（述語動詞），＊語注・文法的留意事項，①②③は訳順番号

(1) 単語や熟語の説明を読みながら，部分訳を書きなさい。わからない単語の意味は調べなさい。
(2) 部分訳が終わったら，①②③の訳順番号を参考に，全文和訳を書きなさい。
　全文和訳が終わったら，Step Check Box □をぬりなさい。

　　　　　　　② 　　　　　①
1) Preparation for Cargo Transfer

- ＊ preparation：〔名詞〕準備
- ＊ cargo：〔名詞〕船荷，積荷
- ＊ prepare：〔他動詞〕
- ＊ transfer：〔名詞〕移動，〔動詞〕～を移す，運ぶ，動かす

タイトル和訳　_____

タイトルがわかると，以下の英文には「積荷を移す」ことに関する具体的な詳細が述べられていると推測できる。

　　　　　　　　　　①
2) [If a ship **becomes** disabled or stranded,]/

- ＊ if：〔接続詞〕もし～ならば
- ＊ disabled：〔形容詞〕航行不能な
- ＊ stranded：〔形容詞〕座礁した
- ＊接続詞のあとには，主語S＋動詞Vが続く。
- ＊ disabled ship：航行不能船
- ＊ stranded ship：座礁船

①部分訳 _____

　　　⑤　　　　　　④　　　　　　　　②　　　　　　　③
it **may be** necessary/ to transfer all or part of her cargo to another vessel./

- ＊ it：仮主語なので「それは」と訳さない。
  真の主語は to 不定詞以下（to transfer all or part of her cargo to another vessel）
- ＊ all or part of her cargo：船荷のすべてあるいは一部
- ＊ transfer A to B：〔他動詞〕AをBへ移す
- ＊船（ship）は she で表すので，her cargo と書く。

64

②③④部分訳 ＿＿＿＿＿＿＿＿＿＿＿＿＿＿＿＿＿＿＿＿＿＿＿＿＿＿＿＿＿＿ ことは必要だろう

全文和訳 ＿＿

3) The operation ⑤ will be greatly ⑥ expedited/

* operation：〔名詞〕作業　　　　* operate：〔他動詞〕～を操作する　　* greatly：〔副詞〕大いに，とても
* will be greatly Ved：〔未来形の受動態〕～られるだろう
* expedite：(仕事などを)手速く処理する(片づける)

⑤⑥部分訳 ＿＿＿＿＿＿＿＿＿＿＿＿＿＿＿＿＿＿＿＿＿＿＿＿＿＿＿＿＿＿＿＿＿＿＿＿＿＿

[if ① the ship to be lightened ④ makes ③ the following preparations/ (at the place) where ② (it is) possible:/

* the ship to be lightened：軽くされるべき船 → 座礁船
* lighten：〔他動詞〕(船などの荷を)軽くする　　　* lightened：過去分詞形なので受け身のように訳す。
* the following preparations：次の準備　(以下の英文 4)～9) は preparations の具体例である)
* where it is possible：可能な場所で　(where は関係副詞で，at the place が省略されている)

①③④部分訳　もし ＿＿＿＿＿＿＿＿＿＿＿＿＿＿＿＿＿＿＿＿＿＿＿＿＿＿＿ ならば

全文和訳 ＿＿

4) ③ Establish ② contact/ with ① the lightening vessel/

* establish：〔他動詞〕～を確立する = set up　(動詞が文頭にきていて，命令文なので「～しなさい」と訳す)
* contact with：～との連絡，交渉
* lightening vessel：軽い船＝貨物を受け取る船 → 瀬取り船
* lightening：〔現在分詞形〕積荷などが軽くなるという意味で，形容詞の働きをする。

①②③部分訳 ＿＿＿＿＿＿＿＿＿＿＿＿＿＿＿＿＿＿＿＿＿＿＿＿＿＿＿＿＿＿＿＿＿＿＿＿

and make ⑧ a detailed plan/ of ⑦ the proposed operation/

* make a plan：計画を立てなさい(命令文)　　　　* detailed：〔形容詞〕詳細な
* proposed：(過去分詞形→受け身)提案された，計画された　　* operation：〔名詞〕業務，作業

⑧部分訳 ＿＿＿＿＿＿＿＿＿＿＿＿＿＿＿＿＿＿＿＿＿＿

including ⑥ the designation of ⑤ a communications ④ channel./

* including：〔前置詞〕～を含めて　　*前置詞 including 以下は，直前の名詞 the proposed operation にかかる。
* designation：〔名詞〕指定　　* communications channel：〔名詞〕通信チャンネル

④⑤⑥⑦部分訳 ＿＿＿＿＿＿＿＿＿＿＿＿＿＿＿＿＿＿＿＿＿＿＿＿＿＿＿＿＿＿ についての

全文和訳 ＿＿

5) ① Fixed or portable hand-held radio telephones may be ② usefully employed./

* fixed or portable hand-held radio telephone：固定のあるいは携帯用の手で持てる無線電話
* usefully：〔副詞〕有益に　　　　　　　　　* useful：〔形容詞〕有効な，役にたつ，助けになる
* employ：〔他動詞〕利用する，用いる，～を使う，雇用する
* may：〔助動詞〕契約書，制定法などの英文で用いられて「～しなければいけない」「～するものとする」と訳す。

全文和訳 ＿＿

6) **Lay out** mooring lines, heaving lines, messengers, stoppers, fenders, etc./
　⑦　　　　①　　　　　②　　　　　③　　　　④　　　⑤　　　⑥

* lay out：〜を用意する　（動詞で始まる文なので，これは命令文である。「①〜⑥を用意しなさい」となる）
* mooring line：〔名詞〕係留索
* heaving line：〔名詞〕ヒービングライン　　　　　* heaving：〔名詞〕（錨などの）引き上げ，投げ出し
* messenger：〔名詞〕補助索　　　　　　　　　　　* fender：〔名詞〕防舷材（フェンダー）
* etc.：et cetera の略　（エトセトラと発音し，「〜など」と訳す）

全文和訳

7) [If no large fenders are available,]/
　　　　①

* if：〔接続詞〕後に主語S＋動詞Vが続く。
* no large fenders：主語が no と打ち消しなので，和訳するときは，動詞を打ち消す（＝ are not available）。
* available：〔形容詞〕入手する，利用できる，暇がある

①部分訳 _____

mooring ropes and any other soft material should be stung/
　　　　　　　　②　　　　　　　　　　⑤

* mooring rope：〔名詞〕係留索　　　　　　　* any other soft material：そのほかの柔らかい材料
* should be stung：助動詞＋受動態なので「〜が急いで用意されるべきである」と訳す。
* sting-stung-stung：〔他動詞〕〜を急いで用意する

②⑤部分訳 _____

over the ship's side in strategic positions./
　　　③　　　　　　　　　　④

* over the ship's side：舷側上方の　　　　　　* in strategic position：効果的な位置に
* over：〔前置詞〕（場所・年齢を）超えて，〜の上に（あるものの上に円弧を描くように覆っているイメージ）

③④和訳 _____

全文和訳

8) **Have** the anchors cleared ready for use,/
　　④　　　　　　　　③

* have + 名詞 + Ved：have は使役動詞で「〜させる」という意味がある。
* clear：〔他動詞〕片づける（障害物がない状態），ここでは錨を不備なく（問題なく）使用に適した状態にすることを指す。
* ready for use：使用準備ができた

③④部分訳 _____

[if (they are) in waters/ where use may be possible.]/
　　①　　　　　　　　　　②

* they are が省略されている。they ＝ the anchors　　　* use：〔名詞〕使用　　* water：〔名詞〕水域
* in waters where use may be possible：関係副詞 where 以下は waters を修飾している。

①②和訳　錨が _____ あるならば

全文和訳

9) **Brief** the officers and crew/ on the operation,/ with particular reference to the safety aspects/
　　④　　　　　　③　　　　　　　　　②　　　　　　　　　　①

* brief：〔他動詞〕〜に要旨を説明する　　　* on the operation：作業について → Brief にかかる
* particular：〔形容詞〕特別な　　　　　　* reference：〔名詞〕参照，言及
* safety：〔名詞〕安全　　　　　　　　　　* safe：〔形容詞〕　　　　* aspect：〔名詞〕局面，様子
* with particular reference to the safety aspects：安全面に特別にふれながら → Brief にかかる

②③④部分訳 _____

and **complete** the appropriate safety check list/ [if (it is) possible.]/
　　⑦　　　　　　　　　　　　　　⑥　　　　　　　　　⑤

＊ complete：〔他動詞〕〜を完成させる　（命令文なので，「④説明しなさい，⑦そして完成させなさい」と訳す）
＊ appropriate：〔形容詞〕適切な　　　　　　　　　　＊ if it is possible：可能ならば

⑥⑦部分訳＿＿＿＿＿＿＿＿＿＿＿＿＿＿＿＿＿＿＿＿＿＿＿＿＿＿＿＿＿

全文和訳

Step 3　網掛けは主語，**ゴシック体**は動詞（述語動詞）

(1) 音声を聞きながら，音読する。音読が終わったら，Step Check Box □をぬりなさい。
(2) 左から順に，スラッシュ毎の意味を考える。

Preparation for Cargo Transfer

　If a ship **becomes** disabled or stranded,/ it **may be** necessary/ to transfer all or part of her cargo to another vessel./　The operation **will be** greatly **expedited**/ if the ship to be lightened **makes** the following preparations/ where possible:/
− **Establish** contact/ with the lightening vessel/ and **make** a detailed plan/ of the proposed operation/ including the designation of a communications channel./　Fixed or portable hand-held radio telephones **may be** usefully **employed**./
− **Lay out** mooring lines, heaving lines, messengers, stoppers, fenders, etc./　If no large fenders **are** available,/ mooring ropes and any other soft material **should be stung**/ over the ship's side in strategic positions./
− **Have** the anchors cleared ready for use,/ if in waters/ where use **may be** possible./
− **Brief** the officers and crew/ on the operation,/ with particular reference to the safety aspects/ and **complete** the appropriate safety check list/ if possible./

【専門英語単語リスト】書いておぼえよう

発音したり，スペル練習したり，意味を調べたりしたらチェック欄の○をぬりなさい。
辞書を引いて単語の意味が多くある場合，英語の文脈から判断し，適切な意味を選びなさい。

英文	単語	品詞	意味を調べて書きなさい	発音しながらスペル練習しなさい	チェック
1)	preparation / prepare	名 / 動			○○○
1)	transfer	名詞			○○○
3)	expedite	他動詞			○○○
3)	lighten	他動詞			○○○
4)	establish	他動詞			○○○
4)	detailed / detail	形 / 動			○○○
4)	propose	他動詞			○○○
4)	designation	名詞			○○○
5)	employ	他動詞			○○○
6)	lay out	熟語			○○○
7)	available	形容詞			○○○
8)	clear	他動詞			○○○
8)	waters	名詞			○○○
9)	brief	他動詞			○○○
9)	complete	他動詞			○○○

Unit 16　積荷の注意事項

Step Check Box □ □ □

Step 1　下記の英文を読み，次の3つの活動をしなさい。終了後 Step Check Box □ をぬりなさい。

(1) 動詞（述語動詞）を○で囲む。動詞の意味を考えて，動詞のあとに続く内容を推測する。
(2) 主語（主部）に下線を引く。
(3) スラッシュを入れる。スラッシュを入れる場所は主に下記の3つ。
- コンマ (,) やピリオド (.) でスラッシュ (/) を入れる。音読するときは，スラッシュで息つぎする。
- 接続詞の前にスラッシュを入れる。［接続詞＋主語＋動詞］は文としてまとまった意味を表す。
  ここでいう接続詞とは when, if, until, before, after など (and, but, or, so を含まない)。
- 原則として前置詞の前にスラッシュを入れる。(前置詞＋名詞) は，最小の意味のまとまりを表す。

1) Stowage
2) Openings in the weather deck over which cargo is stowed shall be securely closed and battened down.
3) The ventilators and air pipes shall be efficiently protected.
4) Timber deck cargoes shall extend over at least the entire available length which is the total length of the well or wells between superstructures.
5) Where there is no limiting superstructure at the after end, the timber shall extend at least to the after end of the aftermost hatchway.
6) The timber deck cargo shall extend athwartships as close as possible to the ship's side, due allowance being made for obstructions such as guard rails, bulwark stays, uprights, pilot access, etc., provided any gap thus created at the side of the ship shall not exceed a mean of 4% of the breadth.
7) The timber shall be stowed as solidly as possible to at least the standard height of the superstructure other than any raised quarter-deck.

出典：2級海技士（航海）平成26年2月定期試験問題

Step 2　網掛けは主語，**ゴシック体**は動詞（述語動詞），＊語注・文法的留意事項，①②③は訳順番号

(1) 単語や熟語の説明を読みながら，部分訳を書きなさい。わからない単語の意味は調べなさい。
(2) 部分訳が終わったら，①②③の訳順番号を参考に，全文和訳を書きなさい。
全文和訳が終わったら，Step Check Box □ をぬりなさい。

1) Stowage

* stowage：〔名詞〕積荷，積み込み　　　　　　* stow：〔他動詞〕しまい込む，きっちり詰める
* タイトルが「積荷」。「安全性を高める積荷」に関する知識があれば，英文を理解しやすい。タイトルに続く英文には，タイトルに関する具体的あるいは詳細な情報が書かれている。
* 原則として，英文では，左から右に読み進めていくと，情報が詳細になる。このことを理解しながら，スラッシュ毎に立ち止まり，スラッシュ毎の意味を考えて読むと，和訳しやすくなる。

タイトル和訳

2) ③ Openings in the weather deck/ ② [over which cargo ① **is stowed**]/ ④ **shall be** ⑤ securely **closed** and ⑥ **battened down**./

* opening：〔名詞〕開口部，通路　　　　　　* weather deck：〔名詞〕暴露甲板
* ①②③の意味は「積荷がきっちり詰められている暴露甲板の開口部は」となる。
* securely：〔副詞〕しっかりと，安全に　　　* shall：〔助動詞〕〜す（される）べきである
* batten down：船倉口を密閉する（ここでは受動態で書かれていて，shall be battened down の意味）
* 接続詞 and は，何と何をつないでいるか？ → 述語動詞 shall be securely closed と 述語動詞 battened down をつなぐ。

全文和訳

3) The ventilators and air pipes **shall be** efficiently **protected**./
   ① ventilators and air pipes ② protected

- * ventilator：〔名詞〕通風設備，換気装置
- * air pipe：通気管
- * efficiently：〔副詞〕能率的に，有効に
- * protect：〔動詞〕= keep safe
- * ventilate：〔他動詞〕換気する，通風する
- * effective：〔形容詞〕効果がある
- * efficient：〔形容詞〕
- * shall be protected：〔受動態〕守られるべきである

全文和訳

4) **Timber deck cargoes** **shall extend**/ over at least the entire available length/ [**which is** the total length of the well or wells between superstructures.]/
   ① Timber deck cargoes ② which is ③ of the well or wells ④ between superstructures ⑤ the entire available length ⑥ shall extend

- * timber：〔名詞〕木材
- * extend：〔自動詞〕（範囲が）ひろがる，わたる
- * at least：〔熟語〕少なくとも
- * entire：〔形容詞〕全体の = whole
- * length：〔名詞〕長さ
- * timber deck cargo：甲板積み木材
- * over：〔前置詞〕〜の一面に，〜を覆った
- * little-less-least：little の比較級，最上級
- * available：〔形容詞〕利用できる
- * long：〔形容詞〕長い
- * well：ウェル（船楼間の甲板（well deck），船舶のポンプを囲んだ区画部）
- * well deck：ウェルデッキ（凹甲板：船首楼と船尾楼の間の甲板）
- * superstructure：上部構造部，船楼（上甲板上の構造物で，船楼や甲板室など）
- * which：〔関係代名詞〕which 以下は at least the entire available length にかかる。which 以下は The entire available length is the total length of the well or wells between superstructures. の意味である。
  ②③④⑤の意味は「ウエルの全長あるいは上部構造部間にあるウエルの，少なくとも利用できる全体の長さ」となる。

①⑥部分訳 _____ ⑤部分訳 _____

全文和訳

5) [**Where there is** no limiting superstructure/ at the after end,]/ the timber **shall extend**/ at least to the after end/ of the aftermost hatchway./
   ① at the after end ② no limiting superstructure ③ Where there is ④ the timber shall extend ⑤ of the aftermost hatchway ⑥ at least to the after end

- * where：〔関係副詞〕where の前に At the place が省略されている。「〜のところでは」と訳す。
- * limiting：〔形容詞〕制限する
- * superstructure：〔名詞〕上部構造部，船楼
- * there is no limiting 〜：名詞を否定しているので，和訳では動詞を打ち消す。
- * at the after end：〔熟語〕船尾に
- * after：〔形容詞〕（船の位置を示す場合）船尾の方にある
- * to the after end：〔熟語〕船尾の方へ
  after の直前には the，after の直後には名詞 end があるので，after は形容詞の働きをしている。
- * to：〔前置詞〕〜の方へ（方向を示す）
- * at：〔前置詞〕〜に，〜で（場所を示す）
- * aftermost：〔形容詞〕船の最後部の
- * hatchway：〔名詞〕（甲板の）昇降口，（あげ蓋のある）出入り口
- * extend：〔自動詞〕（土地，道などが）広がる，までに伸びる（及ぶ）

①②③部分訳 _____ がないところでは

⑤⑥部分訳　（木材は）_____ 少なくとも船尾の方へ，（広げて積むべきである）

全文和訳

6) **The timber deck cargo** **shall extend** athwartships/ as close as possible/ to the ship's side,/
   ① The timber deck cargo ② to the ship's side ③ as close as possible ④ athwartships ⑤ shall extend

- * athwartships：〔副詞〕船体を横切って
- * close：〔副詞〕接近して，近く（発音に注意）
- * to the ship's side：船の舷側の方へ
- * athwart：〔副詞〕船の竜骨と直角に，横に，横断して
- * as 〜 as possible：できるだけ〜

②③④部分訳　（木材は）_____ （広げて積むべきである）

due allowance being made for obstructions /such as guard rails, bulwark stays, uprights, pilot access, etc.,/

* due：〔形容詞〕十分な，相当な
* allowance：〔名詞〕余裕部分，見込みしろ，ゆとり
* being made for ～：〔分詞構文〕これは，〔接続詞＋主語S＋動詞V…〕で書きかえることができる。
  being made for 以下を書きかえると And due allowance shall be made for obstruction such as …となる。
* obstruction：〔名詞〕障害物
* such as：〔熟語〕～のような （障害物の具体例が such as 以下に示される）
  「guard rails 以下のような障害物のために」と訳す。
* guardrail：〔名詞〕手すり，ガードレール
* bulwark stay：〔名詞〕船の舷しょう（船首甲板の舷側に備えられる波除け），ブルワークスティ
* upright：〔名詞〕支柱（舷側において木材を支える）　　* pilot access：〔名詞〕水先人の出入口
* etc.：et cetera の略。「エトセトラ」と発音し，「など」と訳す。

⑪⑫⑬⑭部分訳 _____ 十分な余裕スペースがつくられる

[provided (that) any gap/ thus created at the side of the ship **shall not exceed** a mean of 4% of the breadth.]/

* provided that 主語S＋動詞V…：〔接続詞〕もし that 以下ならば
* provided that ＝ if
* created：〔過去分詞〕つくられる（受身に訳す），gap にかかる。
* at the side of the ship：船の舷側に
* exceed：〔他動詞〕～を超える，～を上回る
* gap：〔名詞〕すきま，空所
* a mean of 4%：平均して4%
* mean：〔名詞〕平均，平均値
* breath：〔名詞〕幅
* broad：〔形容詞〕幅広い

⑥⑦⑧⑨⑩部分訳　もし _____ ならば

全文和訳

7) The timber **shall be stowed**/ as solidly as possible/ to at least the standard height of the superstructure/ other than any raised quarter-deck./

* stow：〔他動詞〕～を積み込む
* shall be stowed：積まれるべきである
* solidly：〔副詞〕固く
* as ～ as possible：できるだけ～
* to：〔前置詞〕～まで （ここでは「少なくとも～の標準の高さまで」の意味）
* height：〔名詞〕高さ
* high：〔形容詞〕
* superstructure：（船，橋脚，船楼などの）上部構造物
* quarter-deck：船尾甲板
* raised quarter-deck：一段高くなった船尾甲板
* raised：〔形容詞〕一段高い，一段高くなった
* other than ～：〔熟語〕～のほかに（＝ besides），～を除いて

②③④部分訳 _____

全文和訳

Step 3 網掛けは主語，**ゴシック体**は動詞（述語動詞）

(1) 音声を聞きながら，音読する。音読が終わったら，Step Check Box □をぬりなさい。
(2) 左から順に，スラッシュ毎の意味を考える。

Stowage
(1) Openings in the weather deck/ over which cargo **is stowed**/ **shall be** securely **closed** and **battened down**./
(2) The ventilators and air pipes **shall be** efficiently **protected**./
(3) Timber deck cargoes **shall extend**/ over at least the entire available length/ which **is** the total length of the well or wells between superstructures./
(4) Where there **is** no limiting superstructure/ at the after end,/ the timber **shall extend**/ at least to the after end/ of the aftermost hatchway./
(5) The timber deck cargo **shall extend** athwartships/ as close as possible/ to the ship's side,/ due allowance being made for obstructions/ such as guard rails, bulwark stays, uprights, pilot access, etc.,/ provided any gap/ thus created at the side of the ship **shall not exceed** a mean of 4% of the breadth./
(6) The timber **shall be stowed**/ as solidly as possible/ to at least the standard height of the superstructure/ other than any raised quarter-deck./

【専門英語単語リスト】書いておぼえよう

発音したり，スペル練習したり，意味を調べたりしたらチェック欄の○をぬりなさい。
辞書を引いて単語の意味が多くある場合，英語の文脈から判断し，適切な意味を選びなさい。

英文	単語	品詞	意味を調べて書きなさい	発音しながらスペル練習しなさい	チェック
1)	stow / stowage	動／名			○○○
2)	weather	名詞			○○○
2)	batten	他動詞			○○○
3)	protect	他動詞			○○○
4)	extend	自動詞			○○○
4)	available	形容詞			○○○
4)	long / length	形／名			○○○
5)	limit	他動詞			○○○
5)	aftermost	形容詞			○○○
5)	hatchway	名詞			○○○
6)	as close as possible	熟語			○○○
6)	obstruction	名詞			○○○
6)	exceed	他動詞			○○○
6)	mean	名詞			○○○
7)	high / height	形／名			○○○

Unit 17　カーゴタンクの原油洗浄　　Step Check Box □ □ □

Step 1　下記の英文を読み，次の3つの活動をしなさい。終了後 Step Check Box □ をぬりなさい。

(1) 動詞（述語動詞）を○で囲む。動詞の意味を考えて，動詞のあとに続く内容を推測する。
(2) 主語（主部）に下線を引く。
(3) スラッシュを入れる。スラッシュを入れる場所は主に下記の3つ。
- コンマ (,) やピリオド (.) でスラッシュ (/) を入れる。音読するときは，スラッシュで息つぎする。
- 接続詞の前にスラッシュを入れる。[接続詞＋主語＋動詞] は文としてまとまった意味を表す。
  ここでいう接続詞とは when, if, until, before, after など (and, but, or, so を含まない)。
- 原則として前置詞の前にスラッシュを入れる。(前置詞＋名詞) は，最小の意味のまとまりを表す。

1) Crude oil washing operations
2) Every oil tanker operating with crude oil washing systems shall be provided with an Operations and Equipment Manual detailing the system and equipment and specifying operational procedures.
3) Such a Manual shall be to the satisfaction of the Administration and shall contain all the information set out in the specifications referred to in paragraph 2 of regulation 33 of this Annex.
4) If an alteration affecting the crude oil washing system is made, the Operations and Equipment Manual shall be revised accordingly.
5) With respect to the ballasting of cargo tanks, sufficient cargo tanks shall be crude oil washed prior to each ballast voyage in order that, taking into account the tanker's trading pattern and expected weather conditions, ballast water is put only into cargo tanks which have been crude oil washed.

出典：2級海技士（航海）平成21年10月 & 平成26年10月定期試験問題
MARPOL 条約附属書I, Chapter 4, Regulation 35 からの抜粋

Step 2　網掛けは主語，**ゴシック体**は動詞（述語動詞），＊語注・文法的留意事項，①②③は訳順番号

(1) 単語や熟語の説明を読みながら，部分訳を書きなさい。わからない単語の意味は調べなさい。
(2) 部分訳が終わったら，①②③の訳順番号を参考に，全文和訳を書きなさい。
　　全文和訳が終わったら，Step Check Box □ をぬりなさい。

1) Crude oil washing operations

- ＊ crude oil：〔名詞〕原油
- ＊ operation：〔名詞〕作業，業務，操作
- ＊ operate：〔自動詞〕〜が作動（稼動）する，仕事をする

タイトル和訳

＊タイトルがわかれば，以下の英文に書かれている内容が，原油洗浄作業について限定されることがわかる。
　原油洗浄作業に関する既知事項を整理しながら，英文を読むと理解しやすい。豊かな専門知識があれば，英文内容を推測しやすい。

2) ③ Every oil tanker/ ② operating with crude oil washing systems ① **shall be provided** with ⑦

- ＊ operate with 〜：〜を扱う
- ＊ 名詞＋ Ving：現在分詞 operating が，直前の名詞 every oil tanker を修飾している。
- ＊ washing system：洗浄装置
- ＊ shall be provided with：〔助動詞＋受動態〕⑥以下を備え付けているべきである

①②③部分訳 _____

72

an Operations and Equipment Manual/detailing the system and equipment/ and specifying operational procedures./

* an Operations and Equipment Manual：装置操作マニュアル，手引書　　　* equipment：〔名詞〕装置，装備
* detail：〔他動詞〕詳述する，～を列挙する
* detailing：〔現在分詞〕直前の名詞 Manual にかかる。「システムと装置を詳述している」と訳す。
* specify：〔他動詞〕～を具体的に述べる
* specifying：〔現在分詞〕直前の名詞 Manual にかかる。「操作手順を具体的に述べている」と訳す。
* operational：〔形容詞〕　　　　　　　　　* operate：〔他動詞〕　　　　　　　* procedure：〔名詞〕手順，順序

④⑤⑥部分訳 _____

全文和訳

3) Such a Manual shall be/ to the satisfaction of the Administration/ and

* manual：〔名詞〕手引書，マニュアル　（Such a Manual で「そのような手引書は」となる）
* satisfaction：〔名詞〕納得，満足　　　　　　　* administration：〔名詞〕主管庁，管海官庁
* to the satisfaction of ～：～にとって申し分のない，～の満足のいくように
* shall：〔助動詞〕～すべし，～すべきである　　　* and：〔接続詞〕shall be と shall contain をつなぐ。

①②部分訳 _____ そして

shall contain all the information/ set out in the specifications/ referred to in paragraph 2 of regulation 33 of this Annex./

* contain：〔他動詞〕～を含む　　　　　　　* set out：〔過去分詞形〕呈示された
* specification：〔名詞〕明細事項
* 名詞＋過去分詞：過去分詞 set out が，直前の名詞 information を修飾している。
* referred to ～：〔過去分詞形〕言及されている，ふれられている
* in paragraph 2：第 2 項に　　　　　　　* regulation 33：規則 33 条
* annex：（名詞）附属書
* 名詞＋過去分詞：過去分詞 referred to が，直前の名詞 specifications を修飾している。

④⑤⑥⑦部分訳 _____（を含むべきである）

全文和訳

4) [If an alteration/ affecting the crude oil washing system is made,]/

* alteration：〔名詞〕変更　　　　　　　　　* alter：〔他動詞〕～を変える ＝ change
* affect：〔他動詞〕～に影響を与える，変化をもたらす
* affecting：〔現在分詞形〕「～している」と進行形のように訳す。
* 名詞＋Ving（現在分詞）の形ならば，現在分詞 affecting が，直前の名詞 alteration を修飾するように訳す。

①②部分訳 _____

the Operations and Equipment Manual shall be revised accordingly./

* revise：〔他動詞〕～を変える，改訂する　　　　* shall be revised：〔受動態〕改訂されるべきである
* accordingly：〔副詞〕それに応じて

全文和訳

5) With respect to the ballasting of cargo tanks,/ sufficient cargo tanks **shall be crude oil washed**/

- ＊ with respect to ～：〔熟語〕～に関しては
- ＊ cargo tank：カーゴタンク
- ＊ shall be crude oil washed：原油洗浄されるべきだ
- ＊ ballasting：〔名詞〕バラスト（脚荷材料）
- ＊ sufficient：〔形容詞〕十分な数・量の ＝ enough

⑩⑪⑫部分訳 _____

prior to each ballast voyage/ [in order that,/ taking into account the tanker's trading pattern and expected weather conditions,/

- ＊ prior to ～：〔熟語〕～に先だち
- ＊ in order that ～：(that 以下になる) ように （that 以下の主語は ⑤ ballast water, 動詞は ⑧ is put である）
- ＊ take into account ～：〔熟語〕(②と③)を考慮に入れる
- ＊ taking into account：taking は現在分詞形で，「～しながら」と訳し，[接続詞＋主語＋動詞]に書き換え可能。
- ＊ tanker's trading pattern：タンカーの運航形態
- ＊ expected：〔過去分詞〕予想される
- ＊ expected weather condition：予想される天候状態

②③④部分訳 _____

ballast water is put only into cargo tanks/ which **have been crude oil washed**./]

- ＊ is put into ～：〔受動態〕（バラスト水が）～に入れられる
- ＊ have been crude oil washed：〔現在完了形の受動態〕原油洗浄された

⑥⑦部分訳 _____

全文和訳 _____

Step 3　網掛けは主語，ゴシック体は動詞（述語動詞）

(1) 音声を聞きながら，音読する。音読が終わったら，Step Check Box □をぬりなさい。
(2) 左から順に，スラッシュ毎の意味を考える。

Crude Oil washing operations

　Every oil tanker/ operating with crude oil washing systems **shall be provided** with an Operations and Equipment Manual/ detailing the system and equipment/ and specifying operational procedures./　Such a Manual **shall be**/ to the satisfaction of the Administration/ and **shall contain** all the information/ set out in the specifications/ referred to in paragraph 2 of regulation 33 of this Annex./　If an alteration/ affecting the crude oil washing system **is made**,/ the Operations and Equipment Manual **shall be revised** accordingly./

　With respect to the ballasting of cargo tanks,/ sufficient cargo tanks **shall be crude oil washed**/ prior to each ballast voyage/ in order that,/ taking into account the tanker's trading pattern and expected weather conditions,/ ballast water **is put** only into cargo tanks/ which **have been crude oil washed**./

【文法のポイント】

接続詞の前，原則として前置詞の前で，スラッシュを入れる。スラッシュは，意味のまとまりを示す。スラッシュ毎に，左から右へ区切りながら訳せば，英文の内容が理解できる。和訳をする場合は，下記のポイントを参考に，日本語らしい語順に直す。

1. ［接続詞＋主語S＋動詞V…］は，動詞Vにかかる。
2. （前置詞＋名詞（もの・コト））は，直前（直後）の名詞や動詞にかかる。
3. ［関係代名詞＋主語S＋動詞V…］は，直前の名詞にかかる。
4. 名詞＋過去分詞（Ved）の語順の場合，過去分詞（Ved）は，直前の名詞にかかる。
5. 名詞＋現在分詞（Ving）の語順の場合，現在分詞（Ving）は，直前の名詞にかかる。

【専門英語単語リスト】書いておぼえよう

発音したり，スペル練習したり，意味を調べたりしたらチェック欄の○をぬりなさい。
辞書を引いて単語の意味が多くある場合，英語の文脈から判断し，適切な意味を選びなさい。

英文	単語	品詞	意味を調べて書きなさい	発音しながらスペル練習しなさい	チェック
1)	crude oil	熟語			○○○
1)	operation / operate	名／動			○○○
2)	specify	他動詞			○○○
2)	operational / operate	形／動			○○○
3)	manual	名詞			○○○
3)	satisfaction	名詞			○○○
3)	contain	他動詞			○○○
3)	refer to	熟語			○○○
3)	regulation	名詞			○○○
4)	revise	他動詞			○○○
5)	with respect to	熟語			○○○
5)	ballast	名／動			○○○
5)	prior to	前置詞			○○○
5)	in order that S + V	熟語	～するように		○○○
5)	take into account	熟語			○○○

～～～ 理解を深めるキーワード「カーゴタンク原油洗浄」 ～～～

　原油や化学液体物などの液体荷物を港で揚げたあと，液体を積んだ船の倉庫のなか（船倉）の側面や底面などに荷物の液体とゴミ（スラッジ）が残る。それらの液体とスラッジが，次の荷物に不純物として混入しないように，船倉の洗浄を行う。洗浄は，次に運ぶ液体貨物と同じ液体を高圧でシャワーのように噴射し，共洗いする。これをカーゴタンクの液体洗浄（荷物が原油の場合，原油洗浄）と呼んでいる。原油などで共洗いする場合，高圧で噴出するときに発生する静電気が，原油からの可燃ガスに発火すると大爆発を起こしかねないため，必ず洗浄中はイナートガス（不活性ガス）を充填しながら作業を行う。

Unit 18　バラスト水

Step Check Box □ □ □

Step 1　下記の英文を読み，次の3つの活動をしなさい。終了後 Step Check Box □ をぬりなさい。

(1) 動詞（述語動詞）を○で囲む。動詞の意味を考えて，動詞のあとに続く内容を推測する。
(2) 主語（主部）に下線を引く。
(3) スラッシュを入れる。スラッシュを入れる場所は主に下記の3つ。
- コンマ (,) やピリオド (.) でスラッシュ (/) を入れる。音読するときは，スラッシュで息つぎする。
- 接続詞の前にスラッシュを入れる。[接続詞＋主語＋動詞] は文としてまとまった意味を表す。
  ここでいう接続詞とは when, if, until, before, after など (and, but, or, so を含まない)。
- 原則として前置詞の前にスラッシュを入れる。(前置詞＋名詞) は，最小の意味のまとまりを表す。

1) Sampling from the Ballast Water Discharge Line

2) The advantage in sampling the biota present in the ballast water discharge line is that this is most likely to accurately represent the concentration of substances and organisms in the actual discharge, which is of primary concern in assessing compliance with the discharge.

3) The disadvantages of this method are that, on most ships, in-line sampling should be carried out in the engine room, where space may be limited, and the handling of water once the samples were concentrated may be impracticable.

4) In order to undertake an accurate measurement on the organism concentration in the ballast water, it is recommended to install an "isokinetic" sampling facility.

5) Isokinetic sampling is intended for the sampling of water mixtures with secondary immiscible phases (i.e. sand or oil) in which there are substantial density differentials.

6) In such conditions, convergence and divergence from sampling ports is of significant concern.

出典：1級海技士（航海）平成26年4月定期試験問題
IMO Marine Enviroment Protection Committee, Resolution MEPC.173(58)
Guidelines for ballast water sampling から抜粋

Step 2　網掛け は主語，**ゴシック体**は動詞（述語動詞），＊語注・文法的留意事項，①②③は訳順番号

(1) 単語や熟語の説明を読みながら，部分訳を書きなさい。わからない単語の意味は調べなさい。
(2) 部分訳が終わったら，①②③の訳順番号を参考に，全文和訳を書きなさい。
　　全文和訳が終わったら，Step Check Box □ をぬりなさい。

1) Sampling from the Ballast Water Discharge Line

＊ discharge：〔名詞〕放出水，排出（量），流失，〔他動詞〕～を出す，排出する
＊ タイトルは，どのような内容が書かれているのか推測するのに役に立つ。

タイトル和訳 _____

2) The advantage in sampling the biota (being) present/ in the ballast water discharge line **is**/
　　　④　　　　　　　③　　　　　　　②　　　　　　　　　①

＊ The advantage in Ving ～ is that S + V：～の利点は that 以下である　　＊ advantage ↔ disadvantage
＊ biota：〔名詞〕生物相，バイオータ
＊ biology, biography などが関連語で，bio- は「命・人生」の意味。
＊ in the ballast water discharge line：バラスト水排出管のなかに
＊ sample：〔他動詞〕～を採取する　　　　　　　　　＊ in sampling：in + Ving（動名詞）→「～するときに」と訳す。
＊ (being) present：〔形容詞〕存在している　　　　　＊①②が biota を修飾している。

①②③④部分訳 _____

that **this is most likely to** accurately represent
　　　⑤　　　　　　⑨　　　　　⑩

* that：〔接続詞〕〜ということ　（that 以下には主語 S と動詞 V がある）
* this：ここでは biota を指す。　　　　　　　　　* likely-more likely-most likely：〔形容詞〕
* be most likely to V：〔熟語〕もっとも〜しそうである
* accurately：〔副詞〕的確に，間違いなく
* represent：〔他動詞〕〜を指摘する，〜をはっきりと述べる　（「何を」指摘するのかは⑥⑦⑧に書いてある）

⑤⑨⑩部分訳 _____

the concentration/ of substances and organisms/ in the actual discharge,/
　　⑧　　　　　　　⑦　　　　　　　　　　⑥

* concentration：〔名詞〕集積　　　* substance：〔名詞〕物質　　　* organism：〔名詞〕有機物
* actual discharge：実際の排出（水）　* in：〔前置詞〕in の基本的な意味は「〜のなかにある」。

⑥⑦⑧部分訳 _____

which is of primary concern/ in assessing compliance with the discharge./
　⑪　　　　　⑭　　　　　　　　　　⑬　　　　　　　　　⑫

* which：which は⑧ the concentration を指している。⑥ discharge の後にカンマがあるので，ここでは「そしてその concentration（集積）は」と訳す。
* primary concern：いちばん懸念されること　　　*英語では左から右へ⑭⑬⑫の順に詳細な情報が述べられる。
* in assessing：in + 動名詞→「〜するときに」と訳す。　* assess：〔他動詞〕評価する

⑪⑫⑬⑭部分訳：そしてその集積は，排出水が法を遵守しているかどうか評価するときに _____ である

全文和訳 _____

3) The disadvantages of this method **are**/ [that,/on most ships,/ **in-line sampling** should be carried out/
　　　　①　　　　　　　　　　　　⑫　　　②　　　　　　　③　　　　　　⑤　　　　　　　　⑥
in the engine room,/
　　④

* disadvantage：不利な点は [that…may be impracticable] ということである。disadvantage ↔ advantage
  1)では利点を述べ，3)では不利な点を述べている。
* this method：このやり方。ここでは in-line sampling を指す。
* in-line sampling：インラインサンプリング（動作中の管から，なかの流体を直接採取する方式）
* should be carried out：実施されているはずだ

③④⑤⑥部分訳 _____（ということである）

where **space may be limited**,/ and **the handling of water**/ once **the samples** were concentrated/
　　　　⑦　　　　　　　　　　　　　　⑩　　　　　　　　　⑧　　　　　　　　⑨
may be impracticable.]/
　⑪

* where：〔関係副詞〕先行詞は the engine room である。ただし where の前（room の後）にカンマがあるので，「そこでは（機関室では），空間が〜」と訳す。
* space：〔名詞〕空間，スペース　　　　　　　* may be limited：制限されているかもしれない
* once：〔接続詞〕いったん（ひとたび）〜すると　* concentrate：〔他動詞〕〜を集める
* once the samples were concentrated：サンプル水がいったん集積されると
* handling：〔名詞〕取扱い，処理
* impracticable：〔形容詞〕実用的ではない，実行不可能な ↔ practicable〔形容詞〕実用的な

全文和訳 _____

77

4) In order to undertake an accurate measurement/ ⑤ on the organism concentration/ ④ in the ballast water,/ ③ ② ①

* in order to V：〔熟語〕〜するために
* undertake：〔他動詞〕〜を保証する
* an accurate measurement：正確な測定
* on：〔前置詞〕〜に関して
* on the organism concentration：バラスト水のなかの有機物集積に関して

it **is recommended**/ ⑦ to install an "isokinetic" sampling facility./ ⑥

* it：仮主語。真の主語は不定詞以下（to install 以下）なので「〜することが」と訳す。
* to install an "isokinetic" sampling facility：等速吸引設備を置くことが

全文和訳

5) Isokinetic sampling **is intended** for the sampling of water mixtures/ with secondary immiscible phases/ ① ⑥ ⑤ ④ ③
(i.e. sand or oil)/ ②

* is intended for：〜が意図されている
* water mixtures：水混合物
* with：〔前置詞〕with は，基本的に「〜と一緒に」という意味。「〜において，〜のとき」と解釈してもよい。
* secondary：〔形容詞〕副次的に
* immiscible：〔形容詞〕非混和性の（混ざらない）⇔ miscible：混和性の
* i.e.：ラテン語 id est (= that is) の略語で，「つまり」の意味である。that is と発音する。
* ②③の大意は「たとえば砂やオイルなどの非混和性の第二相」となる。

[in which there **are** substantial density differentials.]/ ⑦ ⑨ ⑧

* in which：〔関係副詞〕= where。直前の名詞 phases にかかるが，和訳するときは「そしてそこでは〜」と訳す。
* substantial：〔形容詞〕相当な，十分な
* density：〔名詞〕濃度，密度
* differential：〔名詞〕差
* different：〔形容詞〕違って

⑦⑧⑨部分訳　そしてそこでは（第二相では），かなり濃度に差があります

全文和訳

6) In such conditions,/ convergence and divergence from sampling ports **is** of significant concern./ ① ② ③

* convergence：〔名詞〕集中，収束 ⇔ divergence：発散
* port：〔名詞〕（ガス・水などの）出入り口。ここではサンプリングの抽出口のこと。
* of concern：〔熟語〕重大な，関心事の
　of + 名詞は形容詞の働きをする。たとえば，of use = useful，of no use = useless がその例である。
* significant：〔形容詞〕意義深い，かなりの = important
* ②大意：(バラスト水の)抽出口からの収束や発散は

全文和訳

Step 3　網掛け は主語，**ゴシック体**は動詞（述語動詞）

(1) 音声を聞きながら，音読する。音読が終わったら，Step Check Box □をぬりなさい。
(2) 左から順に，スラッシュ毎の意味を考える。

Sampling from the Ballast Water Discharge Line

1　The advantage in sampling the biota present/ in the ballast water discharge line **is**/ that this **is most likely**

to accurately represent the concentration/ of substances and organisms/ in the actual discharge,/ which **is** of primary concern/ in assessing compliance with the discharge./

2 The disadvantages of this method **are**/ that,/ on most ships,/ in-line sampling **should be carried out**/ in the engine room,/ where space **may be limited**,/ and the handling of water/ once the samples **were concentrated**/ **may be** impracticable./

3 In order to undertake an accurate measurement/ on the organism concentration/ in the ballast water,/ it **is recommended**/ to install an "isokinetic" sampling facility./ Isokinetic sampling **is intended** for the sampling of water mixtures/ with secondary immiscible phases/ (i.e. sand or oil)/ in which there **are** substantial density differentials./ In such conditions,/ convergence and divergence from sampling ports **is** of significant concern./

【専門英語単語リスト】書いておぼえよう

発音したり，スペル練習したり，意味を調べたりしたらチェック欄の○をぬりなさい。
辞書を引いて単語の意味が多くある場合，英語の文脈から判断し，適切な意味を選びなさい。

英文	単語	品詞	意味を調べて書きなさい	発音しながらスペル練習しなさい	チェック
1)	discharge	他動詞			○○○
2)	advantage ↔ disadvantage	名詞			○○○
2)	present / absent	形容詞			○○○
2)	be likely to V	熟語			○○○
2)	represent	他動詞			○○○
2)	concentration / concentrate	名／動			○○○
2)	substance	名詞			○○○
2)	organism	名詞			○○○
2)	assess / assessment	動／名			○○○
2)	compliance / comply	名／動			○○○
3)	carry out	熟語			○○○
3)	impracticable ↔ practicable	形容詞			○○○
4)	facility	名詞			○○○
5)	density / dense	名／形			○○○
6)	convergence ↔ divergence	名詞			○○○

～理解を深めるキーワード「カーゴタンク原油洗浄」～

　バラスト水とは，船舶の喫水やトリムを調整するために用いられる水である。船舶が空荷のときに，バラスト水をタンクに積載することで喫水を深くして，船体を安定させるために用いられる。バラスト水は，通常，バラスト水を積載するための専用のバラストタンクに積載される。一方で，バラストタンクを持たない船舶においては，荷下し後の貨物艙にバラスト水が積載される。原油タンカーなどの貨物艙にバラスト水を積載する場合，バラスト水には油分が含まれることとなり，このバラスト水を海に排出することで，海洋汚染を引き起こす原因となる。これを防ぐために，このようなバラスト水を排出するときは，スロップタンク装置や油水分離装置を用いて事前処理を行った上で，油分濃度を監視しながら排出を行う必要がある。また，バラスト水には積載する海域の海水などが用いられるため，バラスト水にはこれを取り入れた海域に生息するプランクトンなどの生物が含まれている。このバラスト水を，採水した海域とは別の海域において排出することで，排出した海域に本来生息していない生物が排出され，本来の生態系が破壊される環境被害が発生している。これを防止するために，バラスト水中の生物が基準以下となるように設備で事前処理を行ってから排出する必要がある。これらの内容は，MARPOL条約（海洋汚染防止条約；International Convention for the Prevention of Pollution from Ships, 1974）や，BWM条約（船舶のバラスト水及び沈殿物の規制および管理のための国際条約；BWM Convention, 2004）に規定されており，国内法では海洋汚染等及び海上災害の防止に関する法律に規定されている。このUnitのバラスト水のサンプリングは，バラスト水管理システムの処理性能が基準を満たしているかを確認するために，バラスト水を採取分析するために用いられるものである。

Unit 19 船員の保護設備

Step Check Box ☐ ☐ ☐

Step 1 下記の英文を読み，次の3つの活動をしなさい。終了後 Step Check Box ☐ をぬりなさい。

(1) 動詞（述語動詞）を○で囲む。動詞の意味を考えて，動詞のあとに続く内容を推測する。
(2) 主語（主部）に下線を引く。
(3) スラッシュを入れる。スラッシュを入れる場所は主に下記の3つ。
- コンマ (,) やピリオド (.) でスラッシュ (/) を入れる。音読するときは，スラッシュで息つぎする。
- 接続詞の前にスラッシュを入れる。[接続詞＋主語＋動詞]は文としてまとまった意味を表す。
  ここでいう接続詞とは when, if, until, before, after など（and, but, or, so を含まない）。
- 原則として前置詞の前にスラッシュを入れる。（前置詞＋名詞）は，最小の意味のまとまりを表す。

1) The deckhouses used for the accommodation of the crew shall be constructed to an acceptable level of strength.
2) Guard-rails or bulwarks shall be fitted around all exposed decks.
3) The height of the bulwarks or guard-rails shall be at least 1m from the deck, provided that where this height would interfere with the normal operation of the ship, a lesser height may be approved, if the Administration is satisfied that adequate protection is provided.
4) Deck cargo carried on any ship shall be so stowed that any opening which is in way of the cargo and which gives access to and from the crew's quarters, the machinery space and all other parts used in the essential operation of the ship can be closed and secured against water ingress.
5) Protection for the crew in the form of guard-rails or lifelines shall be provided above the deck cargo if there is no convenient passage on or below the deck of the ship.

出典：2級海技士（航海）平成22年4月定期試験問題
International Convention on Load Lines（LL条約）附属書Ⅰ Regulation 25 からの抜粋

Step 2 網掛けは主語，ゴシック体は動詞（述語動詞），＊語注・文法的留意事項，①②③は訳順番号

(1) 単語や熟語の説明を読みながら，部分訳を書きなさい。わからない単語の意味は調べなさい。
(2) 部分訳が終わったら，①②③の訳順番号を参考に，全文和訳を書きなさい。
全文和訳が終わったら，Step Check Box ☐ をぬりなさい。

1) ② ① ④
The deckhouses used for the accommodation of the crew **shall be constructed**/

* deckhouse：〔名詞〕甲板室　　　　　　　　　* accommodation：〔名詞〕居住設備，生活の場
* used：名詞＋過去分詞の形。過去分詞 used が直前の名詞 deckhouse にかかり，「〜のために使用される」と訳す。
* shall be constructed：〔受動態〕建設されるべきである

③
to an acceptable level of strength./

* acceptable：〔形容詞〕受け入れられる，許容しうる　　* accept：〔他動詞〕
* strength：〔名詞〕強度，強さ　　　　　　　　　　　　* strong：〔形容詞〕
* to an acceptable level of strength：強度が許容できる程度に
  ③（前置詞＋名詞）は，④動詞にかかる。

全文和訳

2) ① ③ ②
Guard-rails or bulwarks **shall be fitted**/ around all exposed decks./

* guard-rail：〔名詞〕ガードレール，手すり
* bulwark：〔名詞〕ブルワーク，船の舷しょう（船首甲板の舷側に備えられる波よけ）

Unit 19 船員の保護設備

＊shall be fitted：〔受動態〕に備え付けられるべきだ
＊exposed：〔過去分詞形〕受け身のように訳す。　　　　＊exposed deck：〔熟語〕暴露甲板
＊around：〔前置詞〕〜の周囲に，〜のあちこちに，〜を一回りして，〜を巡って

全文和訳

　　　　　　　　　　　　　　⑬　　　　　⑯　　　⑭　　　⑮
3) The height of the bulwarks or guard-rails shall be/ at least 1m from the deck,/

＊height：〔名詞〕高さ　　　　　　　　　　　　＊high：〔形容詞〕
＊at least：〔熟語〕少なくとも　　　　　　　　　＊little-less (lesser) -least：少ない-より少ない-最も少ない

⑭⑮⑯部分訳 _____

　　　　　　　　⑫　　　　　　　④　　　①　　　　　③　　　　　　　　　　②
[provided that/ (at the place) [where this height would interfere with the normal operation of the ship,]/

＊provided that 主語S＋動詞V：provided は接続詞で，if と同じ意味。「もし that 以下という条件ならば」となる。
＊where：〔関係副詞〕where の前に at the place が省略されている。where には「どこ」という意味はない。
　[where＋主語＋動詞] が the place にかかり，「〜の場所では」と訳す。
＊interfere with 〜：〔熟語〕〜に抵触する　　　　＊would：〔助動詞〕（ひょっとすると）〜かもしれない
＊normal operation of the ship：船舶の通常の作業

①②③④部分訳（1mという）高さが _____ 場所では

⑩　　　　　⑪
a lesser height may be approved,/

＊lesser：little〔形容詞〕の比較級。little-lesser-least, little-less-least と2通りの変化形がある。
＊a lesser height：より低い高さ　　　　　　　　＊may：〔助動詞〕〜できる，〜ねばならない
＊may be approved：〔助動詞＋受動態〕承認される
＊⑩⑪⑫大意：より低い高さが承認されるという条件ならば → 少し高さが低くても

⑨　　　　　　　　　　⑤　　　　　　⑧　　　　　　　　　　　⑥　　　　　　⑦
if the Administration is satisfied/ that adequate protection is provided./

＊if：〔接続詞〕〜するときは，もし〜ならば。if の後にS＋Vが続く。
＊that：〔接続詞〕〜ということ
＊administration：〔名詞〕主管庁，管海官庁
＊be satisfied：〜を確信している，（規格などに）〜が合っている，〜が満足している
＊adequate：〔形容詞〕適切な　　＊protection：〔名詞〕防護　　＊is provided：装備されている

⑤⑥⑦⑧⑨部分訳　管海官庁が _____ するときは

全文和訳

　　　　　　　　　　　⑪　　　　　　　　　　　　　⑫
4) Deck cargo carried on any ship shall be so stowed/

＊deck cargo：〔名詞〕甲板貨物
＊carried：〔過去分詞〕運ばれる　　　　　　　　＊carried on any ship は直前の名詞 deck cargo にかかる。
＊stow：〔他動詞〕（積荷や食料）を積みこむ，（帆や各種の船具類）を所定の位置に格納する
＊so 〜 that：〔熟語〕⑩ that 以下になるように…だ

⑩⑪⑫部分訳 _____ 積み込まれるべきだ

that any opening/ [which **is** in way of the cargo/ and which **gives** access to and from the crew's quarters,]/
⑩ ③ ① ②

* opening：〔名詞〕開口部　　　　　　　　　　* (be) in way of ～：～の近くに（ある）
* give access to and from the crew's quarters：乗組員の区域へ行き来する
* which：〔関係代名詞〕which 以下は any opening を修飾する働きをしている。
  ここでは，which 以下が2つあり，2つとも any opening を修飾している。

①②③部分訳 _____

the machinery space and all other parts used in the essential operation of the ship
④ ⑥ ⑤

* the machinery space：機関室　　　　　　　　* all other parts：その他の区画
* used：名詞＋過去分詞の形。過去分詞が直前の名詞（all other parts）を修飾する働きをする。
* in the essential operation of the ship：船舶の必要な作業に（使われる）

④⑤⑥部分訳　開口部や _____ が

can be closed and (**can be**) **secured**/ against water ingress./
⑦ ⑨ ⑧

* can be closed：閉鎖される　　　　　　　　　* secure against：～の危険から守る ＝ protect
* against：〔前置詞〕～を防いで，～に備えて　　* water ingress：浸水
* can be closed と can be secured の主語は③④⑤⑥で，「③や④や⑤⑥が閉鎖され，浸水から守られる」と訳す。

⑦⑧⑨部分訳 _____ ように

全文和訳 _____

5) Protection for the crew/ in the form of guard-rails or lifelines **shall be provided**/
⑦ ⑥ ④ ⑤ ⑨

* protection for：～のための安全確保　　　　　* in the form of ～：～という形態で
* or lifelines：あるいは救命索（という形で）　　* shall be provided：用意されるべきである

⑦⑧部分訳 _____

above the deck cargo/ if there **is** no convenient passage/ on (the deck) or below the deck of the ship./
⑧ ③ ② ①

* above：〔前置詞〕～より上に ↔ deck cargo：甲板積み貨物　　* passage：〔名詞〕通路
* if：〔接続詞〕if 以下には主語 S (no convenient passage)＋動詞 V (is) がある。
* on the deck と below the deck は直前の名詞 convenient passage を修飾している。
* below：〔前置詞〕～より下に ↔ above
* there is no convenient passage：名詞が否定されている（no convenient passage）場合は，動詞（is）を打ち消しの形で訳す。
  ここでは，There is not という意味になる。

①②③部分訳　もし _____

全文和訳 _____

Step 3　網掛け は主語，**ゴシック体**は動詞（述語動詞）

(1) 音声を聞きながら，音読する。音読が終わったら，Step Check Box □をぬりなさい。
(2) 左から順に，スラッシュ毎の意味を考える。

The deckhouses used for the accommodation of the crew **shall be constructed**/ to an acceptable level of strength./ Guard-rails or bulwarks **shall be fitted**/ around all exposed decks./ The height of the bulwarks or guard-rails **shall be**/ at least 1m from the deck,/ provided that/ where this height **would interfere** with the normal operation of the ship,/ a lesser height **may be approved**,/ if the Administration **is satisfied**/ that adequate protection **is provided**./

Deck cargo carried on any ship **shall be** so **stowed**/ that any opening/ which **is** in way of the cargo/ and which **gives** access to and from the crew's quarters,/ the machinery space and all other parts used in the essential operation of the ship **can be closed** and **secured**/ against water ingress./ Protection for the crew/ in the form of guard-rails or lifelines **shall be provided**/ above the deck cargo/ if there **is** no convenient passage/ on or below the deck of the ship./

【文法のポイント】
接続詞の前，原則として前置詞の前で，スラッシュを入れる。スラッシュは，意味のまとまりを示す。
スラッシュ毎に，左から右へ区切りながら訳せば，英文の内容が理解できる。和訳をする場合は，下記のポイントを参考に，日本語らしい語順に直す。

1. ［接続詞＋主語S＋動詞V…］は，動詞Vにかかる。
2. (前置詞＋名詞（もの・コト))は，直前（直後）の名詞や動詞にかかる。
3. ［関係代名詞＋主語S＋動詞V…］は，直前の名詞にかかる。
4. 名詞＋過去分詞(Ved)の語順の場合，過去分詞(Ved)は，直前の名詞にかかる。
5. 名詞＋現在分詞(Ving)の語順の場合，現在分詞(Ving)は，直前の名詞にかかる。

【専門英語単語リスト】書いておぼえよう

発音したり，スペル練習したり，意味を調べたりしたらチェック欄の○をぬりなさい。
辞書を引いて単語の意味が多くある場合，英語の文脈から判断し，適切な意味を選びなさい。

英文	単語	品詞	意味を調べて書きなさい	発音しながらスペル練習しなさい	チェック
1)	accommodation	名詞			○○○
1)	construct	他動詞			○○○
2)	fit	他動詞			○○○
2)	exposed deck	熟語			○○○
3)	at least	熟語			○○○
3)	provided that S V = if	熟語			○○○
3)	interfere with	熟語			○○○
3)	approve	他動詞			○○○
3)	be satisfied (with)	熟語			○○○
4)	in way of	熟語			○○○
4)	machinery	名詞			○○○
4)	secure	他動詞			○○○
4)	ingress	名詞			○○○
5)	protection	名詞			○○○
5)	deck cargo	熟語			○○○

Unit 20　条約の適用

Step Check Box □ □ □

Step 1　下記の英文を読み，次の3つの活動をしなさい。終了後 Step Check Box □ をぬりなさい。

(1) 動詞（述語動詞）を○で囲む。動詞の意味を考えて，動詞のあとに続く内容を推測する。
(2) 主語（主部）に下線を引く。
(3) スラッシュを入れる。スラッシュを入れる場所は主に下記の3つ。
- コンマ (,) やピリオド (.) でスラッシュ (/) を入れる。音読するときは，スラッシュで息つぎする。
- 接続詞の前にスラッシュを入れる。［接続詞＋主語＋動詞］は文としてまとまった意味を表す。
  ここでいう接続詞とは when, if, until, before, after など（and, but, or, so を含まない）。
- 原則として前置詞の前にスラッシュを入れる。（前置詞＋名詞）は，最小の意味のまとまりを表す。

1) Ships when engaged on international voyages between the near neighboring ports of two or more States may be exempted by the Administration from the provisions of the present Convention, so long as they shall remain engaged on such voyages, if the Governments of the States in which such ports are situated shall be satisfied that the sheltered nature or conditions of such voyages between such ports make it unreasonable or impracticable to apply the provisions of the present Convention to ships engaged on such voyages.

2) The Administration may exempt any ship which embodies features of a novel kind from any of the provisions of this Convention the application of which might seriously impede research into the development of such features and their incorporation in ships engages on international voyages.

3) Any such ship shall, however, comply with safety requirements which, in the opinion of that Administration, are adequate for the service for which it is intended and are such as to ensure the overall safety of the ship and which are acceptable to the Government of the States to be visited by the ship.

出典：1級海技士（航海）平成24年2月定期試験問題
International Convention on Load Lines（LL条約）Article 6 からの抜粋

Step 2　網掛け は主語，**ゴシック体**は動詞（述語動詞），＊語注・文法的留意事項，①②③は訳順番号

(1) 単語や熟語の説明を読みながら，部分訳を書きなさい。わからない単語の意味は調べなさい。
(2) 部分訳が終わったら，①②③の訳順番号を参考に，全文和訳を書きなさい。
　全文和訳が終わったら，Step Check Box □ をぬりなさい。

1-1) Ships [when (they **are**) **engaged** on international voyages/
　　　　　①　　　　④　　　　　　　③
between the near neighboring ports/ of two or more States/]
　　　　　　　　　　　②

- ＊ they are：省略されているので，補って考えるとわかりやすい。they は，ships を指す。
- ＊ be engaged on：〔熟語〕〜に従事している
- ＊ ③④の大意：国際航海に従事しているときは
- ＊ Ships の動詞は⑲ may be exempted である。
- ＊ neighboring：〔形容詞〕隣接した，同一地域内の
- ＊ State：〔名詞〕（主権を有する）国家，国
- ＊ ①②の大意：2か国あるいはそれ以上の国々にある隣接した港間で

①②③④部分訳　船舶が ＿＿＿＿＿＿＿＿＿＿＿＿＿＿＿＿＿＿＿＿＿＿＿＿ のとき

1-2) **may be exempted**/ by the Administration/ from the provisions of the present Convention,/
　　　　⑲　　　　　　　　　⑰　　　　　　　　　　　　　　⑱

- ＊ exempt：〔他動詞〕（義務・責任などを）免除する
- ＊ may：〔助動詞〕〜するものとする，〜できる，〜ねばならない
- ＊ may be exempted by…from 〜：〔受動態〕主語は 1-1) の Ships で，訳は「（船舶は）…により〜を免除される」となる。
- ＊ administration：〔名詞〕主管庁
- ＊ provision：〔名詞〕規定，条項
- ＊ the present Convention：現在の国際条約
- ＊ ⑰⑱⑲の大意：下記の 1-3) である限り，1-4), 1-5), 1-6) の条件ならば（〜のとき），1-1) 国際航海中の船は，1-2) 主管庁により現在の国際条約の規定を免除される

84

1-3) so long as they **shall remain** engaged on such voyages,/
（⑦　　　⑥　　　　　⑤）

* so long as 主語 S ＋動詞 V：〔接続詞〕〜の限り
* remain：〔自動詞〕〜のままである
* they ＝ ships
* engaged：〔形容詞〕航海中で，従事して

⑤⑥⑦部分訳　船舶が _____

1-4) if the Governments of the States/ [in which such ports **are situated**]/ **shall be satisfied**/
（⑯　　　　⑨　　　　　　　　　　　　　　⑧　　　　　　　　　　⑮）

* in which：〔関係代名詞〕直前の名詞 the States にかかる。
  such ports are situated in the States と言い換えることができる。
* are situated in 〜：〜に置かれている
* the Government：〔名詞〕（集合的に）政府
* shall be satisfied［that S ＋ V］：［that 以下を］確信（納得）するものとする
* 1-4) の文のあと 1-6) まで「政府が何を確信する（be satisfied）のかという内容」が続く。そして，1-4) の文は，1-2) へと意味がつながる。

⑧⑨⑮部分訳　（もし）_____

1-5) [that the sheltered nature or conditions of such voyages between such ports **make** it unreasonable
　　　　　　　　　　　　　　　⑩　　　　　　　　　　　　　　　　　　　　　　　⑫　　⑬
　　or impracticable/
　　⑭

* sheltered：〔形容詞〕（風雨・危険などから）保護されている
* nature：〔名詞〕自然，性質
* condition：〔名詞〕状況，事情
* unreasonable：〔形容詞〕不合理な ↔ reasonable
* impracticable：〔形容詞〕実行不可能な ↔ practicable
* make it unreasonable or impracticable：it は仮目的語なので「それを」と訳さない。it は，1-6) の to apply the provisions …such voyages を指す。make は使役動詞で「（it を）〜にする」という意味。
* ⑩⑫⑬⑭⑮⑯大意：保護されている自然あるいはそのような港間を航行中の状況が（⑪ to apply 以下）するのを不合理で実行不可能にさせていると（政府が）確信するならば

1-6) to apply the provisions of the present Convention to ships engaged on such voyages.]/
　　　　　　　　　　　　　　⑪

* apply A to B：A を B に適用する
* ⑪大意：現在の国際条約の規定を航行中の船舶に適用する

全文和訳

2) The Administration **may exempt** any ship [which **embodies** features of a novel kind]/
　　　①　　　　　　　　⑤　　　　　③　　　　　　　②
　　from any of the provisions of this Convention,/
　　④

* administration：〔名詞〕主管庁，管海官庁
* provision：〔名詞〕規定，条項
* Convention：〔名詞〕国際条約
* exempt A from B：〔他動詞〕A に B を免除する
* embody：〔他動詞〕包含する，持っている
* features of a novel kind：新式の性能
* which 以下は ship を修飾。
* ①②③④⑤大意：主管庁は，新式の性能を持った船舶にこの国際条約の規定を免除できる

the application of which **might** seriously **impede** research/ into the development of such features/
　　　　⑥　　　　　　　　　⑨　　　　　　　　⑧　　　　　　　　　　　　　⑦

* application：〔名詞〕適用
* impede：〔他動詞〕〜を遅らせる，妨げる
* seriously：〔副詞〕ひどく，相当に，本気で
* feature：〔名詞〕特徴，性能
* this Convention, the application of which might seriously impede 〜：which（関係代名詞）は Convention を指す。
  次のように言い換えることができる。→ the application of this Convention might seriously impede 〜
* the application の前にカンマ（,）が付いているので，the application 以下は，理由を付随的に述べている文だとわかる。
* ⑥⑦⑧⑨大意：国際条約の適用が，そのような性能を開発・調査するのを非常に遅らせる

and (the application of this Convention might impede) their incorporation in ships⑫/ engaged⑪ on international⑩ voyages./

* incorporation：〔名詞〕取り込み，合併，合体
* their ＝ features
* impede の目的語は⑦⑧と⑩⑪⑫の2つである。
* ⑩⑪⑫大意：国際航行に従事している船舶にそのような性能を取り込むのを（遅らせる）

全文和訳

3) Any such ship⑬ shall,/ however,/ comply with① safety requirements⑭ [which⑥,/ in the opinion of that Administration②,/ are⑤ adequate④ for the service/ for which it③ is intended/

* any such ship：いかなる船も
* comply with：〔熟語〕規則に従う，条件を満たす
* safety requirement：安全要件（その具体例が⑦⑧⑨⑩⑪⑫に続く）
* in the opinion of that Administration：主管庁の見解により
* be adequate for：〔形容詞〕に適している
* be intended：〔受動態〕予定される
* which are adequate for the service for which it is intended は⑥にかかる。it は any such ship を指す。

②③④⑤⑥部分訳　（どの船舶も）主管庁の見解により予定している航行に適している安全要件に

and (safety requirements)⑧ [(which) are such as to ensure the overall safety of the ship⑦/

* are such as to ensure the overall safety of the ship：船舶の全体的な安全を保証するのに十分なほどの（安全要件に従うべきである）
* (which)：関係代名詞 which が省略されていて，直前に (safety requirements) が略されている。
* such as to V：〔熟語〕〜するようなもの

⑦⑧部分訳　また _____ （安全要件に）

and (safety requirements)⑫ which are acceptable⑪ to the Governments of the States⑩/ to be visited by the ship⑨.]

* the Governments of the States to be visited by the ship：船舶が訪問する国々の政府
* to be visited by the ship：直前の名詞 States にかかる。
* be acceptable to 〜：〔形容詞〕承諾できる，容認できる
* 接続詞 and は何と何をつないでいるか？→⑥，⑧，⑫の safety requirements をつなぐ。
  Any such ship shall, however, comply with が⑥，⑧，⑫に続いている。
* ⑨⑩⑪⑫大意：船舶が訪問する国々の政府にとり承諾できる（安全要件に）

全文和訳

Step 3　網掛けは主語，ゴシック体は動詞（述語動詞）

(1) 音声を聞きながら，音読する。音読が終わったら，Step Check Box □をぬりなさい。
(2) 左から順に，スラッシュ毎の意味を考える。

Ships when **engaged** on international voyages/ between the near neighboring ports/ of two or more States/ **may be exempted**/ by the Administration/ from the provisions of the present Convention,/ so long as they **shall remain** engaged on such voyages,/ if the Governments of the States/ in which such ports **are situated**/ **shall be satisfied**/ that the sheltered nature or conditions of such voyages between such ports **make** it unreasonable or impracticable/ to apply the provisions of the present Convention to ships engaged on such voyages./

The Administration **may exempt** any ship which **embodies** features of a novel kind/ from any of the provisions of this Convention,/ the application of which **might** seriously **impede** research/ into the development of such features/ and their incorporation in ships/ engaged on international voyages./ Any such ship **shall**,/ however,/ **comply with** safety requirements/ which,/ in the opinion of that Administration,/ **are** adequate for the service/ for which it **is intended**/ and **are** such as to ensure the overall safety of the ship/ and which **are** acceptable to the Government of the States/ to be visited by the ship./

【専門英語単語リスト】書いておぼえよう

発音したり，スペル練習したり，意味を調べたりしたらチェック欄の○をぬりなさい。
辞書を引いて単語の意味が多くある場合，英語の文脈から判断し，適切な意味を選びなさい。

英文	単語	品詞	意味を調べて書きなさい	発音しながらスペル練習しなさい	チェック
1)	be engaged on	熟語			○○○
1)	neighbor / neighbour	動詞			○○○
1)	exempt	他動詞			○○○
1)	administration	名詞			○○○
1)	provision	名詞			○○○
1)	so long as	接続詞			○○○
1)	be satisfied (with)	熟語			○○○
2)	embody	他動詞			○○○
2)	feature	名詞			○○○
2)	impede	他動詞			○○○
2)	development / develop	名 / 動			○○○
2)	incorporation	名詞			○○○
3)	comply with	熟語			○○○
3)	adequate	形容詞			○○○
3)	acceptable / accept	形 / 動			○○○

コラム1

述語動詞の見つけ方

1. be 動詞の現在形と過去形 (am, is, are, was, were) は述語動詞で，意味は数学の＝と同じ
2. 助動詞 (will, would, can, could, may, might, must, shall, should) ＋動詞の原形は述語動詞
3. 完了形（have ＋過去分詞）は述語動詞
4. 上のいずれの場合も be 動詞が出てきたら、その後に
　　現在分詞（~ing）があれば（be 動詞＋~ing）のセットで進行形（~しているところ）
　　過去分詞（~ed）があれば（be 動詞＋~ed）のセットで受動態（~されている）

上の 4 つのルールを覚えておけば，少しくらい知らない単語があっても述語動詞は簡単に見つかります。

普通の英文を読む場合は，過去形と過去分詞形が同じ形の動詞の場合に，どちらなのかを判断するのが悩ましいのですが，海技士の英語には過去形があまり使われないので，過去形か過去分詞形かで迷ったら過去分詞形だと考えて読めばよいでしょう。

では実際に英文を読んでみましょう。
次の英文の述語動詞はどれかわかりますか？

Radio transmissions should be made as soon as possible but other means, e.g. rockets and hand flares, should be conserved until it is known that they may attract the attention of aircraft or ships in the vicinity.

上の 4 つのルールを使えば，述語動詞が下のゴシック体の部分だと直ぐにわかりますね。

Radio transmissions **should be made** as soon as possible but other means, e.g. rockets and hand flares, **should be conserved** until it **is known** that they **may attract** the attention of aircraft or ships in the vicinity.

ちなみに述語動詞が見つかれば，主語はその前にあるので，主語も見つけやすくなります。

Radio transmissions **should be made** as soon as possible but other means, e.g. rockets and hand flares, **should be conserved** until it **is known** that they **may attract** the attention of aircraft or ships in the vicinity.

CHAPTER 2

海技士問題 機関

Unit 21 主機空気制御

Step Check Box □ □ □

Step 1　下記の英文を読み，次の3つの活動をしなさい。終了後 Step Check Box □ をぬりなさい。

(1) 動詞（述語動詞）を○で囲む。動詞の意味を考えて，動詞のあとに続く内容を推測する。
(2) 主語（主部）に下線を引く。
(3) スラッシュを入れる。スラッシュを入れる場所は主に下記の3つ。
- コンマ(,)やピリオド(.)でスラッシュ(/)を入れる。音読するときは，スラッシュで息つぎする。
- 接続詞の前にスラッシュを入れる。〔接続詞＋主語＋動詞〕は文としてまとまった意味を表す。
  ここでいう接続詞とは when, if, until, before, after など（and, but, or, so を含まない）。
- 原則として前置詞の前にスラッシュを入れる。〔前置詞＋名詞〕は，最小の意味のまとまりを表す。

1) The operation of the main engine is controlled by means of several pneumatic switching valves, solenoid switching valves, and limit switches even when the main engine is operate by local control.
2) Malfunction of the equipment immediately causes difficulties for engine operation, and at the worst, the main engine may become inoperative.
3) Therefore, it is required that functions and purposes of the equipment used for local control be studied at other times on board the vessel.
4) The actual fitting locations of the components and the locations in the drawings should be also checked.

出典：2級海技士（機関）平成22年10月定期試験問題

Step 2　網掛けは主語，ゴシック体は動詞（述語動詞），＊語注・文法的留意事項，①②③は訳順番号

(1) 単語や熟語の説明を読みながら，部分訳を書きなさい。わからない単語の意味は調べなさい。
(2) 部分訳が終わったら，①②③の訳順番号を参考に，全文和訳を書きなさい。
　全文和訳が終わったら，Step Check Box □ をぬりなさい。

1) The operation of the main engine① is controlled⑦ / by means of⑥ several pneumatic switching valves,③ /

* operation：〔名詞〕
* main engine：主機関
* operate：〔他動詞〕～を操作する
* is controlled：〔受動態〕制御する→「制御される」と訳す。
* 主語は「メインエンジンの操作」，動詞が「制御される」なので，以下に「エンジン制御」関連の情報が書かれている。
* by means of ～：～の手段によって
* several：〔形容詞〕いくつかの（3～6くらいまでの数を指す）
* pneumatic：〔形容詞〕圧縮空気で動く
* several pneumatic switching valves：いくつかの空気切換弁
* by means of 以下には，制御される手段が3つ述べられている。

③④⑤部分訳 ＿＿＿＿＿＿＿＿＿＿＿＿＿＿＿＿＿＿＿＿＿＿＿＿＿＿＿＿＿＿＿＿＿＿＿＿＿＿

solenoid switching valves,④ / and limit switches,⑤ / [even when the main engine is operated② / by local control.] /

* solenoid switching valve：電磁切換え弁（電流が流れると金属棒が動く仕掛けの制御スイッチで，自動制御システムなどで使用）
* limit switch：（エレベーターなどの，ある設定点を通過すると働く）自動制御スイッチ
* even when：〔接続詞〕～のときでも
* local control：機側制御
* 接続詞以下には必ず主語Sと動詞Vがくる。

②部分訳 ＿＿＿

全文和訳 ＿＿

2) Malfunction of the equipment① immediately **causes**③ difficulties/ for engine operation,②/

- * malfunction：〔名詞〕故障，不調，機能不全
- * equipment：〔名詞〕装置
- * cause：〔他動詞〕～を引き起こす，の原因となる
- * operation：〔名詞〕操作
- * operative：〔形容詞〕仕事の，働いて，効力のある
- * inoperative：〔形容詞〕操縦不能の，正常に動かない，作動しない ↔ operative
- * for：〔前置詞〕～のために
  基本的には「～に向かって」というイメージがある。for two hours のように，時間の範囲を示す。また，Thank you for your kindness のように，原因・理由（の範囲）を示す。
- * function：〔名詞〕機能，作用，働く
- * immediately：〔副詞〕ただちに ＝ very soon
- * difficulty：〔名詞〕問題，困難，難事
- * operate：〔他動詞〕～を操作する
- * 単語は，派生語や反対語もファミリー単語（つながる単語）として，ひとくくりにして覚える

①②③部分訳 _____

and at the worst,④/ the main engine⑤ **may become**⑦ inoperative.⑥/

- * at the worst：〔熟語〕最悪の場合は
- * bad（ill）-worse-worst
- * may become：〔助動詞〕なるかもしれない

④⑤⑥⑦部分訳 _____

全文和訳 _____

3) Therefore,/ it① **is required**⑨/

- * therefore：〔副詞〕したがって
- * is required：〔受動態〕必要とする→「必要とされる」と訳す。
- * it は仮主語で，that 以下が真の主語である。「それは」と訳さない。

①⑨部分訳 _____ that 以下が _____

[that functions and purposes of the equipment④⑤/ used for local control③ **(should) be studied**②⑧/ at other times⑦/ on board the vessel.⑥]/

- * that：〔接続詞〕～ということは
  接続詞 that の後には，主語 S（functions and purposes of the equipment used for local control）と動詞 V（should be studied）が続く。
- * function：〔名詞〕機能 * purpose：〔名詞〕目的 * equipment：〔名詞〕装置
- * used：〔過去分詞〕使われる * used for local control は前の名詞 the equipment を修飾している。
- * should be studied：〔受動態〕調査されるべきである。should は省略されている。
- * at other times：他のときに * on board：〔前置詞〕～に乗って * on board the vessel：乗船中の

②③④⑤部分訳 _____ ということは

全文和訳 _____

4) The actual fitting locations of the components and the locations① in the drawings② **should be** also **checked**.③/

- * actual：〔形容詞〕実際の * fitting：〔名詞〕取り付け * location：〔名詞〕位置
- * component：〔名詞〕部品，構成部分 * drawing：〔名詞〕図面 * in：〔前置詞〕～のなかの
- * should be also checked：〔受動態〕調査する，検査する → 調査されねばならない

全文和訳 _____

Step 3   網掛けは主語，**ゴシック体**は動詞（述語動詞）

(1) 音声を聞きながら，音読する。音読が終わったら，Step Check Box □をぬりなさい。
(2) 左から順に，スラッシュ毎の意味を考える。

The operation of the main engine **is controlled**/ by means of several pneumatic switching valves,/ solenoid switching valves,/ and limit switches,/ even when the main engine **is operated**/ by local control./ Malfunction of the equipment immediately **causes** difficulties/ for engine operation,/ and at the worst,/ the main engine **may become** inoperative./ Therefore,/ it **is required**/ that functions and purposes of the equipment/ used for local control **be studied**/ at other times/ on board the vessel./ The actual fitting locations of the components and the locations in the drawings **should be** also **checked**./

【文法のポイント】
接続詞の前，前置詞の前で，スラッシュを入れる。スラッシュは，意味のまとまりを示す。スラッシュ毎に，左から右へ区切りながら訳せば，英文の内容が理解できる。和訳をする場合は，下記のポイントを参考に，日本語らしい語順に直す。
1. ［接続詞＋主語S＋動詞V…］は，動詞Vにかかる。
2. （前置詞＋名詞（もの・コト））は，直前（直後）の名詞や動詞にかかる。
3. ［関係代名詞＋主語S＋動詞V…］は，直前の名詞にかかる。
4. 名詞＋過去分詞（Ved）の語順の場合，過去分詞（Ved）は，直前の名詞にかかる。
5. 名詞＋現在分詞（Ving）の語順の場合，現在分詞（Ving）は，直前の名詞にかかる。

【専門英語単語リスト】書いておぼえよう

発音したり，スペル練習したり，意味を調べたりしたらチェック欄の○をぬりなさい。
辞書を引いて単語の意味が多くある場合，英語の文脈から判断し，適切な意味を選びなさい。

英文	単語	品詞	意味を調べて書きなさい	発音しながらスペル練習しなさい	チェック
1)	operate / operation	動／名			○○○
1)	control / controlled	動／形			○○○
1)	switch	動／名			○○○
1)	limit / limitation	動／名			○○○
2)	malfunction ↔ function	名詞			○○○
2)	equipment / equip	名／動			○○○
2)	bad-worse-worst	形容詞			○○○
2)	inoperative ↔ operative	形容詞			○○○
3)	require / requirement	動／名			○○○
3)	purpose	名詞			○○○
3)	on board	熟語			○○○
3)	vessel	名詞			○○○
4)	fit-fitted-fitted-fitting	他動詞			○○○
4)	component	名詞			○○○
4)	location / locate	名／動			○○○

理解を深めるキーワード「主機の制御」

主機関の制御方式には電気-空気式，空気-空気式，電気-油圧式などの方式があるが，最近よく使われるのは電気-空気式である。これは電気の信号を途中で空気圧に変換し，空気圧の有無でデジタルの"1""0"に相当する論理回路を組み立てたり，空気圧を直にピストンなどに作用させ機器の動作を行わせたりするものである。下図に空気式の記号例，簡単な切替え弁の例，簡略化した起動空気制御の例を示す。

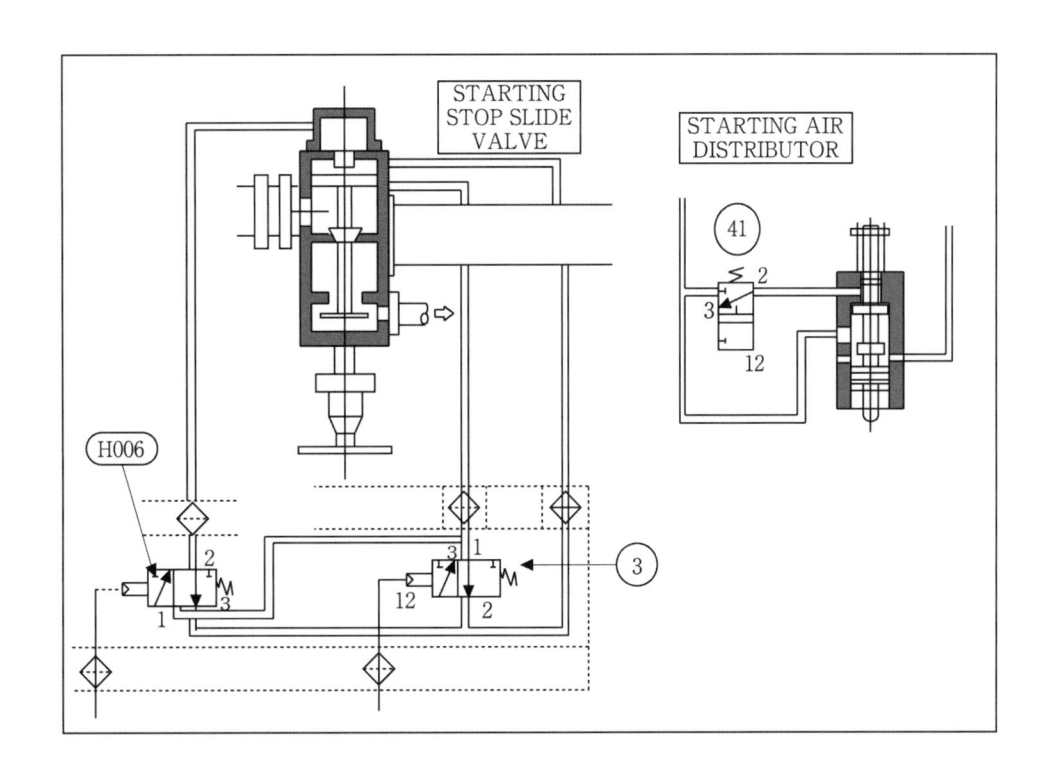

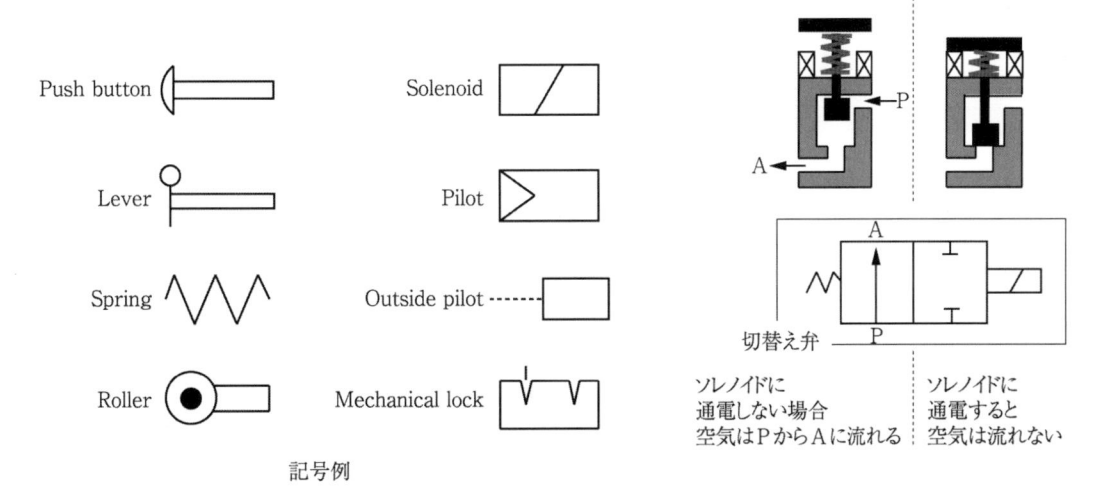

記号例

切替え弁

ソレノイドに通電しない場合 空気はPからAに流れる

ソレノイドに通電すると 空気は流れない

Unit 22　カム・カム軸

Step Check Box □ □ □

Step 1　下記の英文を読み，次の3つの活動をしなさい。終了後 Step Check Box □をぬりなさい。

(1) 動詞（述語動詞）を○で囲む。動詞の意味を考えて，動詞のあとに続く内容を推測する。
(2) 主語（主部）に下線を引く。
(3) スラッシュを入れる。スラッシュを入れる場所は主に下記の3つ。
- コンマ(,)やピリオド(.)でスラッシュ(/)を入れる。音読するときは，スラッシュで息つぎする。
- 接続詞の前にスラッシュを入れる。［接続詞＋主語＋動詞］は文としてまとまった意味を表す。
  ここでいう接続詞とは when, if, until, before, after など（and, but, or, so を含まない）。
- 原則として前置詞の前にスラッシュを入れる。（前置詞＋名詞）は，最小の意味のまとまりを表す。

1) A cam is an eccentric projection on a revolving disk used for the opening and closing of a valve through various intermediate parts, as described above.
2) Originally cams were made as separate pieces and fastened to the camshaft.
3) However, at present most diesel engines, even some in larger sizes, have cams forged or cast integral, meaning as one piece, with the camshaft and then machined, usually ground to the required exact shape.
4) Thus diesel-engine camshafts are similar to automobile-engine camshafts.
5) The advantage of such an integral camshaft is that if one valve of one cylinder is timed correctly, all the valves in all cylinders will be timed correctly.
6) On the other hand, any change in timing will affect all valves and cylinders.

出典：2級海技士（機関）平成22年7月定期試験問題

Step 2　網掛けは主語，ゴシック体は動詞（述語動詞），＊語注・文法的留意事項，①②③は訳順番号

(1) 単語や熟語の説明を読みながら，部分訳を書きなさい。わからない単語の意味は調べなさい。
(2) 部分訳が終わったら，①②③の訳順番号を参考に，全文和訳を書きなさい。
　全文和訳が終わったら，Step Check Box □をぬりなさい。

1)　①A cam **is** an eccentric projection/ ⑥on a revolving disk/ ⑤used for the opening and closing of a valve/ ④
- ＊cam：〔名詞〕カム（回転軸に取り付けて回転運動を上下・前後などの直線運動に変える装置）
- ＊an eccentric projection：〔名詞〕偏心突起（軸や支えが中心を外れた突起）
- ＊revolving disk：回転している円盤　　　　　　　　＊revolve：〔自動詞〕～が回転する
- ＊used：〔過去分詞〕使われる。訳すときは受動態（受け身）のように訳す。
- ＊過去分詞以下（used for the opening and closing of a valve）が名詞（a revolving disk）を修飾している。
- ＊opening and closing：開閉　　　　　　　　　　　　＊valve：〔名詞〕弁，バルブ
- ＊④大意：バルブの開閉のために使用される
- ＊主語と動詞の意味がわかると，1）以下の内容を予想することができる。
　　ここではカム軸の機能や構造の専門知識が必要になる。

⑤⑥部分訳　回転している円盤に付いている＿＿＿＿＿＿＿＿＿＿＿＿＿＿＿＿＿＿＿

through various intermediate parts,/ ③ ②as (it is) described above./
- ＊through：〔前置詞〕～を通って，介して　　＊intermediate：〔形容詞〕中間の　　＊part：〔名詞〕部品
- ＊as (it is) described above：〔熟語〕上記に述べられているように（「上記」の内容は記載されていない）
- ＊③は，④の a valve を修飾する。

③部分訳　＿＿＿＿＿＿＿＿＿＿＿＿＿＿＿＿＿＿＿＿＿＿＿＿＿

全文和訳　＿＿

2) Originally/ cams **were made**/ as separate pieces/ and **(were) fastened** to the camshaft./
① ② ④ ③ ⑥ ⑤

* originally：〔副詞〕もともとは　　　　　　* original：〔形容詞〕最初の，独自の　　　　　* originality：〔名詞〕
* were made：〔受動態〕つくられた　　　　　* as separate pieces：別個の部品として　　　　* separate：〔形容詞〕
* were fastened to：〔受動態〕～に固定された　　　　　　　　　　* camshaft：〔名詞〕カムシャフト（カム軸）
* fasten：〔他動詞〕～をしっかりと固定する，留める　　　　　* fastener：〔名詞〕

全文和訳

3) However,/ at present/ most diesel engines,/ even some in larger sizes,/
① ② ③

* at present：〔熟語〕現在では　　　　　　　* most diesel engine：ほとんどのディーゼル機関
* some (diesel engines) in larger sizes：大きいサイズのディーゼル機関でさえ

②③部分訳

have cams **forged**/ or **cast** integral, meaning as one piece,/ with the camshaft/
⑥ ⑦ ⑤ ④

* have ＋目的語＋Ved（過去分詞）：（目的語）を…される。haveは使役動詞の意味を持つ。
* forged/cast/machined/ground：すべて過去分詞なので「…される」と訳す。　* integral：〔形容詞〕一体化して
* forge：〔他動詞〕～を鍛造する　　　　　　　　　　　　　* cast-cast-cast：〔他動詞〕～を鋳造する
* meaning as one piece：1つの部品として一体化して働くように　　* with：〔前置詞〕～とともに

④⑤⑥⑦部分訳

and then (most diesel engines **have** cams) **machined**,/ usually **ground**/ to the required exact shape./
⑧ ⑩ ⑨

* most diesel engines have cams が省略されている。　　　* machined：〔過去分詞〕機械加工される
* grind-ground-ground：〔過去分詞〕磨かれる
* require：〔他動詞〕必要とする，求める　　　　　　　　* required：〔過去分詞〕受動態（受け身）のように訳す。
* exact：〔形容詞〕正確な，的確な　　　　　　　　　　　* to the required exact shape：要求された正確な形になるまで

⑧⑨⑩部分訳

全文和訳

4) Thus diesel-engine camshafts **are** similar to automobile-engine camshafts./
① ③ ②

* thus：〔副詞〕このように　　　　　　　　　* be similar to 名詞：〔形容詞〕～と類似している
* automobile：〔名詞〕自動車（＝motor car）

全文和訳

5) The advantage of such an integral camshaft **is**/ [that if one valve of one cylinder **is timed** correctly,/
② ① ⑧ ③ ⑤ ④

* advantage：〔名詞〕利点　　　　　　　　　　* such an integral camshaft：そのような一体化したカム軸
* The advantage ＋ is that 主語S＋動詞V：「～の利点は…ということです」という意味。
　that 以下にはカムシャフトの利点について書いてある。専門知識を働かせると英文理解が早い。
* that：〔接続詞〕～ということ→that以下には必ず主語＋動詞が続く。
　この英文では，that以下の主語Sは⑥ all the valves in all cylinders，動詞Vは⑦ will be timed である。
* if：〔接続詞〕もし～ならば
　if の後には，主語 one valve of one cylinder と動詞 is timed が続く。
　if one valve of one cylinder is timed correctly は「もし～ならば」という条件文。⑦ will be timed にかかる。
* is timed：〔受動態〕調整される　　　　　　　* time：〔他動詞〕（装置などのタイミング）を調整する
* 長文和訳では，番号順ではなく，左から順にスラッシュ毎に訳した後に，日本語らしい語順に直すとよい。

①②部分訳　　　　　　　　　　　　　　　　　　③④⑤部分訳

⑥　　　　　　　　　　　　　⑦
all the valves in all cylinders **will be timed** correctly.]/

* in：〔前置詞〕～のなかにある
* correctly：〔副詞〕
* will be timed：未来形の受動態
* correct：〔形容詞〕正確な，正しい

全文和訳

① ② ④ ③
6) On the other hand,/ any change in timing **will affect** all valves and cylinders./

* on the other hand：〔熟語〕一方では
* any change in timing：タイミング調整におけるどんな変化も
* affect：〔他動詞〕に作用する，影響を与える

全文和訳

Step 3　網掛けは主語，**ゴシック体**は動詞（述語動詞）

(1) 音声を聞きながら，音読する。音読が終わったら，Step Check Box □ をぬりなさい。
(2) 左から順に，スラッシュ毎の意味を考える。

　A cam **is** an eccentric projection/ on a revolving disk/ used for the opening and closing of a valve/ through various intermediate parts,/ as described above./　Originally/ cams **were made**/ as separate pieces/ and **fastened** to the camshaft./　However,/ at present/ most diesel engines,/ even some in larger sizes,/ **have** cams **forged**/ or **cast** integral, meaning as one piece,/ with the camshaft/ and then **machined**,/ usually **ground**/ to the required exact shape./　Thus diesel-engine camshafts **are** similar to automobile-engine camshafts./　The advantage of such an integral camshaft **is**/ that if one valve of one cylinder **is timed** correctly,/ all the valves in all cylinders **will be timed** correctly./　On the other hand,/ any change in timing **will affect** all valves and cylinders./

【専門英語単語リスト】書いておぼえよう

発音したり，スペル練習したり，意味を調べたりしたらチェック欄の○をぬりなさい。
辞書を引いて単語の意味が多くある場合，英語の文脈から判断し，適切な意味を選びなさい。

英文	単語	品詞	意味を調べて書きなさい	発音しながらスペル練習しなさい	チェック
1)	cam / camshaft	名詞			○○○
1)	revolve / revolving	自動詞			○○○
1)	valve	名詞			○○○
1)	describe	他動詞			○○○
2)	separate	形 / 動			○○○
2)	fasten	他動詞			○○○
3)	at present	熟語			○○○
3)	forge	他動詞			○○○
3)	cast-cast-cast	他動詞			○○○
3)	machine	動 / 名			○○○
3)	grind-ground-ground	他動詞			○○○
3)	require / required	他動詞			○○○
4)	be similar to 名詞	熟語			○○○
5)	time	他動詞			○○○
6)	affect	他動詞			○○○

理解を深めるキーワード「カム(Cam)について」

　カムは機械部品の運動方向を変える重要な要素の一つである。エンジンのなかではたとえば給・排気弁，燃料噴射ポンプや起動弁駆動用に用いられている。カムにはカム軸と一体となった一体型(a)と，分割してカム軸に固定する分割型(b)がある。一体型の場合，弁などの駆動タイミングを全シリンダー同時に調整することができて便利であるが，個別のシリンダーで対応することができない。分割型の場合は①②のボルトの締め加減によりカム③がカム軸面に沿って回転し，個別にタイミングを調整することができる。

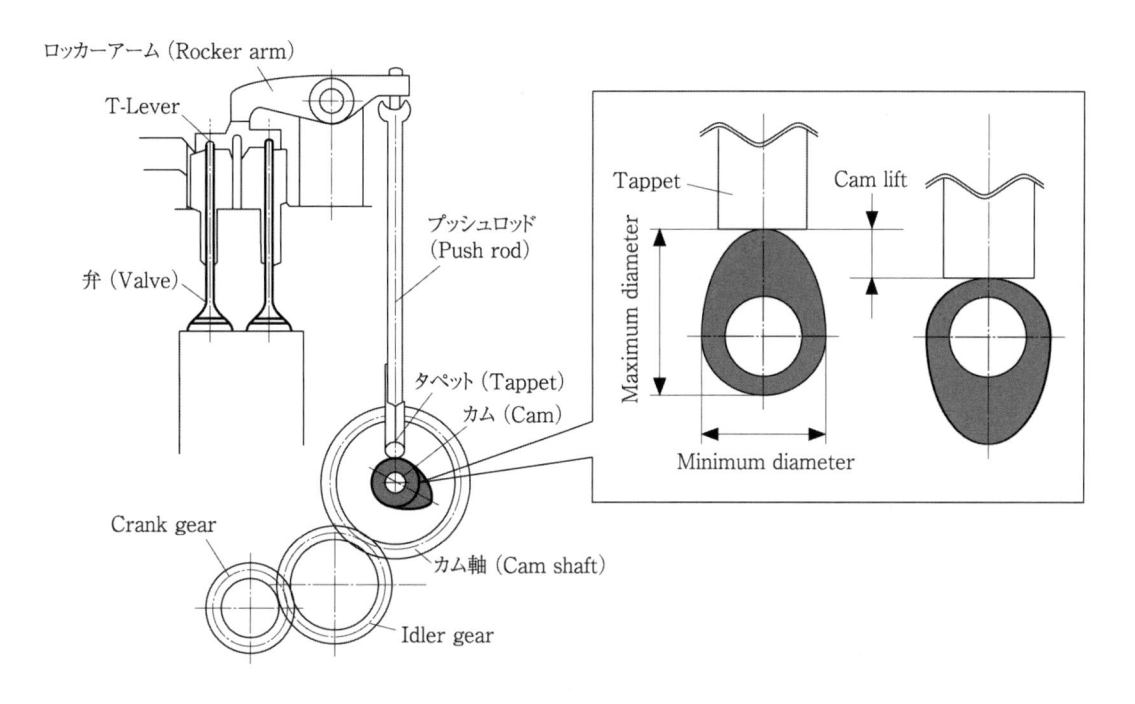

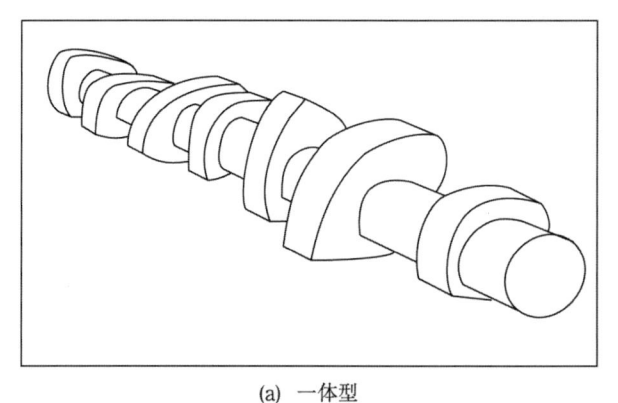

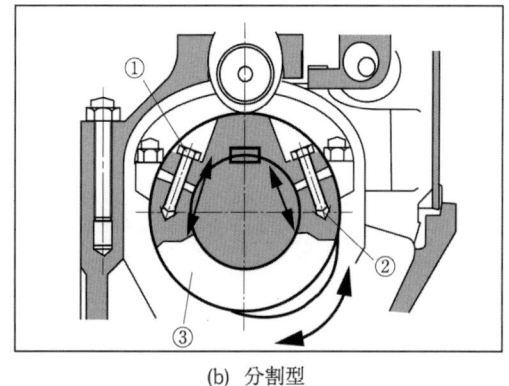

(a) 一体型

(b) 分割型

Unit 23　排気弁

Step Check Box □ □ □

Step 1　下記の英文を読み，次の3つの活動をしなさい。終了後 Step Check Box □をぬりなさい。

(1) 動詞（述語動詞）を○で囲む。動詞の意味を考えて，動詞のあとに続く内容を推測する。
(2) 主語（主部）に下線を引く。
(3) スラッシュを入れる。スラッシュを入れる場所は主に下記の3つ。
- コンマ(,)やピリオド(.)でスラッシュ(/)を入れる。音読するときは，スラッシュで息つぎする。
- 接続詞の前にスラッシュを入れる。[接続詞＋主語＋動詞]は文としてまとまった意味を表す。
  ここでいう接続詞とは when, if, until, before, after など（and, but, or, so を含まない）。
- 原則として前置詞の前にスラッシュを入れる。（前置詞＋名詞）は，最小の意味のまとまりを表す。

1) The exhaust valve consists of a valve housing and a valve spindle.
2) The valve housing is of cast iron and arranged for water cooling.
3) The housing is provided with a bottom piece of steel with Stellite welded onto the seat.
4) The spindle is made of heat resistant steel with Stellite welded onto the seat.
5) The housing is provided with a spindle guide.
6) The exhaust valve housing is connected to the cylinder cover with studs and nuts tightened by hydraulic jacks.
7) The exhaust valve is opened hydraulically and closed by a set of helical springs.

出典：2級海技士（機関）平成23年7月定期試験問題

Step 2　網掛けは主語，ゴシック体は動詞（述語動詞），＊語注・文法的留意事項，①②③は訳順番号

(1) 単語や熟語の説明を読みながら，部分訳を書きなさい。わからない単語の意味は調べなさい。
(2) 部分訳が終わったら，①②③の訳順番号を参考に，全文和訳を書きなさい。
　全文和訳が終わったら，Step Check Box □をぬりなさい。

1) <u>The exhaust valve</u>① **consists of**④ a valve housing② and a valve spindle.③ /

- ＊ exhaust valve：排気弁
- ＊ consist of：〔動詞〕～で構成されている，～から成り立つ = be made of
- ＊ valve housing：弁箱　　　　＊ housing：〔名詞〕外被（機器を囲む外側の構造）
- ＊ valve spindle：弁棒　　　　＊ spindle：〔名詞〕形，寸法が精密で変形量が少ないことが必要な短い回転軸
- ＊ 排気弁（the exhaust valve）が何で構成されているか（consist of = be made of）が，話題の中心である。
  実習あるいは専門科目でしっかり勉強することが，必須の対策である。

全文和訳

2) <u>The valve housing</u>① **is (made) of**③ cast iron②/ and **(is) arranged for**⑤ water cooling.④ /

- ＊ be made of：〔熟語〕～でつくられる
- ＊ cast iron：〔名詞〕鋳鉄
- ＊ be arranged for：〔熟語〕～が準備されている，整えられている = be provided with
- ＊ water cooling：水冷，冷却水

全文和訳

3) <u>The housing</u>① **is provided with**⑤ a bottom piece (made) of steel④/ with Stellite welded③/ onto the seat.② /

- ＊ housing：〔名詞〕弁箱　　　　　　＊ be provided with：〔熟語〕～を備え付けている，～が用意されている
- ＊ a bottom piece：〔名詞〕底部　　　＊ seat：〔名詞〕弁座
- ＊ onto：〔前置詞〕～に，～の上へ　　＊ a bottom piece (made) of steel：鉄でつくられた底部＝鉄製の底部

98

Unit 23 排気弁

* weld：〔他動詞〕（〜を…に）溶接（鍛接）する
* Stellite welded onto the seat：welded は過去分詞で「溶接された」の意味。直前の名詞（Stellite）にかかる。「弁座に溶接されたステライト＝弁座にステライトを溶接した」となる。
* Stellite：ステライト。米国製のコバルト・クロム・炭素・タングステン・モリブデンの合金の総称。とても丈夫で硬く，鋳型やコーティングに用いられる。
* steel with Stellite welded on the seat：弁座にステライトを溶接した鉄

①④⑤部分訳 _____

全文和訳 _____

4) <u>The spindle</u> **is made of** heat resistant steel/ with Stellite welded/ onto the seat./
   　　①　　　　　⑤　　　　　　　　④　　　　　　　　③　　　　②

* spindle：〔名詞〕主軸，心棒，スピンドル
* heat resistant steel：〔名詞〕耐熱鋼
* be made of：〔受け身〕〜で構成されている，から成り立つ
* with Stellite welded：ステライトを溶接した

全文和訳 _____

5) <u>The housing</u> **is provided with** a spindle guide./
   　　①　　　　　③　　　　　　②

* be provided with：〔熟語〕〜が用意されている
* spindle guide：弁棒案内部
* 「動詞」（be provided with, be arranged for, be made of, consist of）は，「主語」がどうなっているかを文章から読み取り訳す。

全文和訳 _____

6) <u>The exhaust valve housing</u> **is connected to** the cylinder cover/ with studs and nuts/ tightened by hydraulic jacks./
   　　①　　　　　　　　　　⑤　　　　　　　　④　　　　　　　　③　　　　　　　　②

* exhaust valve housing：排気弁箱
* exhaust：〔他動詞〕排気する，〔名詞〕排気，排出
* be connected to：〔受動態〕〜に接続している
* cylinder cover：シリンダカバー，シリンダヘッド
* studs：〔名詞〕植込みボルト，スタッド（栓，ねじ，ピンなどの小突起）
* nuts：〔名詞〕ナット，親（留め）ねじ（nut と，はまりあうのが bolt）
* with：〔前置詞〕〜を使って
* tightened：〔過去分詞形〕受け身のように訳し，前の名詞 studs and nuts にかかる。
* tighten：〔他動詞〕〜をしっかり締める，堅くする
* tight：〔形容詞〕きつい
* hydraulic jack：〔名詞〕油圧ジャッキ
* hydro-：「水の，液体の，流体の」という意味
* hydraulically：〔副詞〕油圧的に

②③部分訳 _____

全文和訳 _____

7) <u>The exhaust valve</u> **is opened** hydraulically/ and **(is) closed**/ by a set of helical springs./
   　　①　　　　　③　　　　②　　　　　　　⑤　　　　　　④

* hydraulically：〔副詞〕油圧で
* a set of helical spring：ひと組のコイルバネ（つるまきバネ）

全文和訳 _____

Step 3   網掛けは主語，**ゴシック体**は動詞（述語動詞）

(1) 音声を聞きながら，音読する。音読が終わったら，Step Check Box □をぬりなさい。
(2) 左から順に，スラッシュ毎の意味を考える。

The exhaust valve **consists of** a valve housing and a valve spindle./ The valve housing **is of** cast iron/ and **arranged for** water cooling./ The housing **is provided with** a bottom piece of steel/ with Stellite welded/ onto the seat./ The spindle **is made of** heat resistant steel/ with Stellite welded/ onto the seat./ The housing **is provided with** a spindle guide./ The exhaust valve housing **is connected to** the cylinder cover/ with studs and nuts/ tightened by hydraulic jacks./ The exhaust valve **is opened** hydraulically/ and **closed**/ by a set of helical springs./

【文法のポイント】
◆（前置詞＋名詞（もの・コト））は，直前の名詞や動詞にかかる

例1：the cylinder cover/ (with studs and nuts)/ tightened by hydraulic jacks
- with：〔前置詞〕～といっしょに，さまざまなものとのつながりを表す。
  with ＋ 道具・材料　→　～を使って　（例）with a pen：ペンを使って
  with ＋ 原因　　　　→　～が原因で　（例）stay in bed with a fever：熱で寝ている

◆名詞＋過去分詞（Ved）の語順の場合，過去分詞（Ved）は「～された」と受け身のように訳し，直前の名詞にかかる

例2：a bottom piece/ (made of steel)/ with Stellite (welded/ onto the seat)
- welded は過去分詞形で，直前の名詞 Stellite にかかる（修飾している）形容詞の働きをする。
  「溶接されたステライト」と，ステライトが溶接されている状態を表現している。
- made of steel の made は過去分詞形で，直前の名詞（piece）にかかる形容詞の働きをする。
  「鉄でつくられた（底部）＝鉄製の（底部）」という意味。

【専門英語単語リスト】書いておぼえよう

発音したり，スペル練習したり，意味を調べたりしたらチェック欄の○をぬりなさい。
辞書を引いて単語の意味が多くある場合，英語の文脈から判断し，適切な意味を選びなさい。

英文	単語	品詞	意味を調べて書きなさい	発音しながらスペル練習しなさい	チェック
1)	exhaust valve	名詞			○○○
1)	consist of 名詞	自動詞			○○○
1)	valve housing	名詞			○○○
1)	valve spindle	名詞			○○○
2)	be arranged for	受／熟			○○○
2)	arrange / arrangement	動／名			○○○
3)	water cooling	名詞			○○○
3)	be provided with	受／熟			○○○
3)	weld-welded-welded	他動詞			○○○
4)	be made of	受／熟			○○○
4)	heat resistant steel	名詞			○○○
6)	connect	他動詞			○○○
6)	stud / nut	名詞			○○○
6)	tighten / tight	動／形			○○○
7)	a set of	熟語			○○○

理解を深めるキーワード「排気弁」

　排気弁は弁箱(Valve housing)と弁棒(Valve spindle)から成り，弁箱は植え込みボルト(Stud bolt)により強固にシリンダカバーに据え付けられている。また，弁箱は冷却水により冷却され，その底部にはSeat ringと呼ばれるリングが取り付けられ，リングの弁座面ならびに弁棒弁座面には高温ガスによる吹き抜けを防止するためステライト盛りが行われている。

　弁棒は駆動装置(Actuator)上部のシリンダに作動油(Working oil)により圧力を加えることにより開き，圧力を抜くことによりスプリングの力が勝り，弁を閉じることになる(コイルバネ式)。最近では，このスプリングに替わり空気圧(Air chamberに一定の圧力を加えてスプリングの役目を成す)を使い，弁を押し上げて閉じる方式(空気バネ式)が主流となっている。

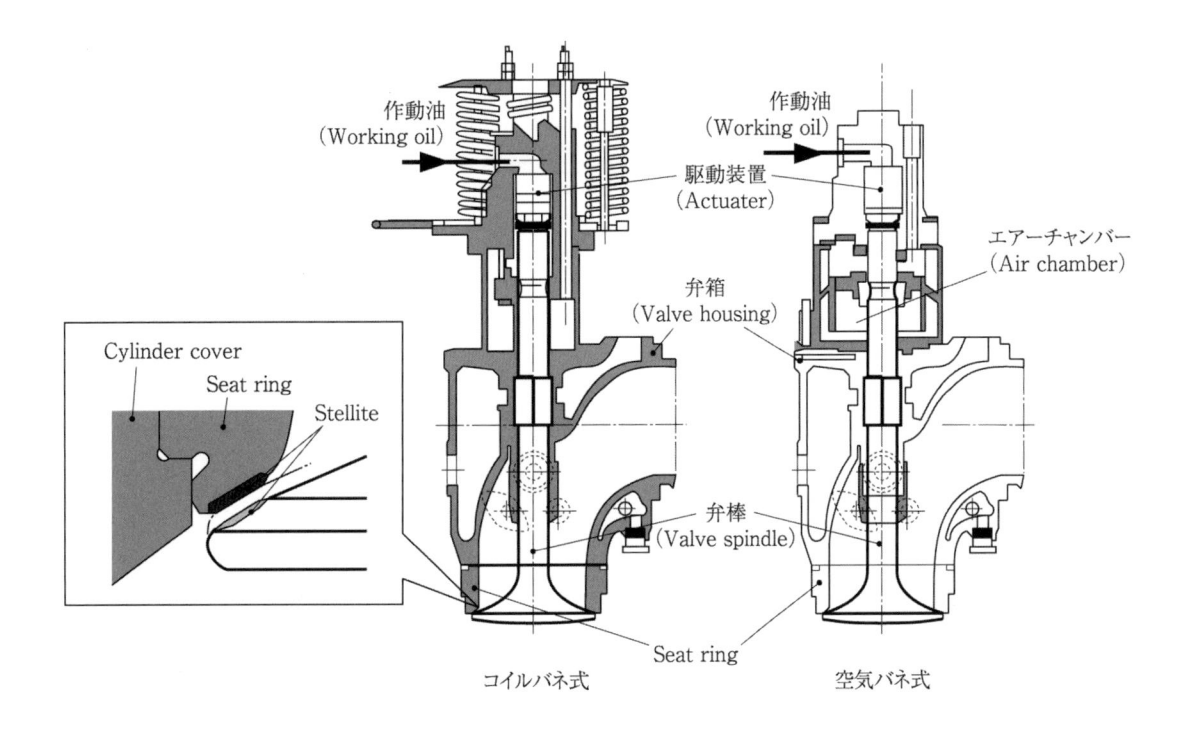

コイルバネ式　　　空気バネ式

Unit 24　ターボチャージャー

Step Check Box ☐ ☐ ☐

Step 1　下記の英文を読み，次の3つの活動をしなさい。終了後 Step Check Box ☐ をぬりなさい。

(1) 動詞（述語動詞）を○で囲む。動詞の意味を考えて，動詞のあとに続く内容を推測する。
(2) 主語（主部）に下線を引く。
(3) スラッシュを入れる。スラッシュを入れる場所は主に下記の3つ。
- コンマ(,)やピリオド(.)でスラッシュ(/)を入れる。音読するときは，スラッシュで息つぎする。
- 接続詞の前にスラッシュを入れる。［接続詞＋主語＋動詞］は文としてまとまった意味を表す。
 ここでいう接続詞とは when, if, until, before, after など（and, but, or, so を含まない）。
- 原則として前置詞の前にスラッシュを入れる。（前置詞＋名詞）は，最小の意味のまとまりを表す。

1) While a vessel was under way, the turbocharger made an abnormal sound and an alarm went off indicating that the inlet exhaust gas temperature in the turbocharger was too high in the diesel generator engine in use.
2) Crewmembers started a standby diesel generator engine and stopped the abnormal diesel generator engine.
3) The turbocharger was overhauled for inspection, and the snap ring for setting the floating bearing of the turbine rotor was discovered to have fallen off;
4) further, scratches on the blower impeller and the casing were discovered.
5) This accident caused deformation of the snap ring groove on the turbine rotor and abnormal abrasions on the floating bearing.

出典：2級海技士（機関）平成23年10月定期試験問題

Step 2　網掛け は主語，ゴシック体は動詞（述語動詞），＊語注・文法的留意事項，①②③は訳順番号

(1) 単語や熟語の説明を読みながら，部分訳を書きなさい。わからない単語の意味は調べなさい。
(2) 部分訳が終わったら，①②③の訳順番号を参考に，全文和訳を書きなさい。
 全文和訳が終わったら，Step Check Box ☐ をぬりなさい。

1) [While a vessel **was** under way,]①/ the turbocharger **made** an abnormal sound②/ and an alarm **went off**④ indicating③/

* while：〔接続詞〕～している間　　　　　　　　＊ be under way：〔熟語〕（船が）航行中である
* turbocharger：〔名詞〕排気タービン過給機（排気を利用したタービンでシリンダ内に空気を強制的に圧送する）
* abnormal：〔形容詞〕異常な ↔ normal：〔形容詞〕ふつうの，自然な，正常な
* 大意は「～の間，排気タービン過給機が異常な音をたてる，すると次に～」となる。
* alarm：〔名詞〕警告音　　　　　　　　　　　　＊ go off：〔熟語〕（警報などが）鳴り出す
* [indicating that 主語S＋動詞V]は分詞構文といい，直前の動詞 went off にかかる。if it indicated that S＋V と同じ意味。
* indicate：〔他動詞〕～を示す（ならば）
* ⑤⑪⑫の大意：that 以下であることが示されるとき，警報が鳴り出す

①②③④部分訳＿＿＿＿＿＿＿＿＿＿＿＿＿＿＿＿＿＿＿＿＿＿＿＿＿＿＿＿＿＿＿＿＿＿＿＿＿＿＿

[that the inlet exhaust gas temperature⑦/ in the turbocharger⑥ **was** too high⑩/ in the diesel generator engine⑨/ in use.⑧]/

* that 主語S＋動詞V：〔接続詞〕～ということを
* inlet：〔名詞〕入り口，注入口　　　　　　　　＊ exhaust gas temperature：排気ガス温度
* in the diesel generator engine：ディーゼル発電機関のなかでは
* in use：使用中の（engine にかかる）

⑥⑦⑧⑨⑩部分訳＿＿＿＿＿＿＿＿＿＿＿＿＿＿＿＿＿＿＿＿＿＿＿＿＿＿＿＿＿＿＿＿＿＿＿＿＿

全文和訳＿＿＿

2) Crewmembers **started** a standby diesel generator engine/ and **stopped** the abnormal diesel generator engine./
   ① ③ ② ⑤ ④

   * crewmember：〔名詞〕乗組員
   * standby：〔形容詞〕控えの，予備の

   全文和訳

3) The turbocharger **was overhauled**/ for inspection,/ and the snap ring/ for setting the floating bearing
   ① ③ ② ⑦ ⑥ ⑤
   of the turbine rotor **was discovered**/ to have fallen off;/
   ④ ⑨ ⑧

   * turbocharger：〔名詞〕ターボチャージャー，排気タービン過給機
   * overhaul：〔他動詞〕必要な修理を施す，整備する，分解する   * for inspection：点検（検査）のために
   * snap ring：止め輪   * floating：〔形容詞〕浮いている，浮動の
   * for setting the floating bearing：浮動軸を固定するための
   * the turbine rotor：タービンロータ，タービン翼車
   * turbine：〔名詞〕タービン（羽車を回転し動力を伝える機械）
   * rotor：回転子
   * fall off：〔動詞〕（離れて）落ちる，（質が）低下する，脱落する   * fall-fell-fallen
   * be discovered to have fallen off：to have + Ved は完了不定詞で，過去の事実を指す。「すでに～だったことが発見される」となる。

   ①②③部分訳 _____

   ④⑤⑥⑦部分訳 _____

   全文和訳

4) further,/ scratches on the blower impeller and the casing **were discovered**./
   ① ④ ② ③ ⑤

   * further：〔副詞〕さらに   * scratch：〔名詞〕かき傷，がりがりという音
   * blower：〔名詞〕（内燃機関の）過給機，送風機，送風装置，ブロワー
   * impeller：〔名詞〕（渦巻きポンプ，扇風機などの）翼車，羽車
   * casing：〔名詞〕囲い，（機械）ケーシング（タービンやポンプで機械内部を密閉する部分）

   全文和訳

5) This accident **caused** deformation of the snap ring groove/ on the turbine rotor/
   ① ⑤ ④ ③ ②
   and (**caused**) abnormal abrasions/ on the floating bearing./
   ⑦ ⑥

   * this accident：前の 3) や 4) を指す。
   * cause：〔他動詞〕～を引き起こす，生じさせる   * deformation：〔名詞〕変形
   * groove：〔名詞〕細長いくぼみ，溝   * the snap ring groove：止め輪溝
   * on the turbine rotor：前置詞 on は，原則として「～の上に付いている，～の上にある」の意味
   * abrasion：〔名詞〕摩耗，擦過傷   * the floating bearing：浮動軸受

   全文和訳

Step 3　網掛けは主語，**ゴシック体**は動詞（述語動詞）

(1) 音声を聞きながら，音読する。音読が終わったら，Step Check Box □をぬりなさい。
(2) 左から順に，スラッシュ毎の意味を考える。

　While a vessel **was** under way,/ the turbocharger **made** an abnormal sound/ and an alarm **went off** indicating/ that the inlet exhaust gas temperature/ in the turbocharger **was** too high/ in the diesel generator engine/ in use./ Crewmembers **started** a standby diesel generator engine/ and **stopped** the abnormal diesel generator engine./

　The turbocharger **was overhauled**/ for inspection,/ and the snap ring/ for setting the floating bearing of the turbine rotor **was discovered**/ to have fallen off;/ further,/ scratches on the blower impeller and the casing **were discovered**./　This accident **caused** deformation of the snap ring groove/ on the turbine rotor/ and abnormal abrasions/ on the floating bearing./

【合格のための重要ポイント】
- 海技士試験は，辞書を持ち込むことができる。日ごろの英語の勉強でも，面倒がらずに，辞書を引く練習をする。
- 海技士試験の英文では，be + Ved（＝受動態）が多く使用されているので，動詞を見つけやすい。
- 動詞がわかると，動詞の左側に置かれている主語「〜は」「〜が」を見つけやすい。
- 最初の英文の主語が，書かれている英文内容のトピックに関連している。たとえば，最初の英文は the turbocharger made an abnormal sound である。だから，「排気タービン過給機の異常音」がトピックであることがわかる。
- 専門知識が豊かであれば，英文を理解しやすい。

【専門英語単語リスト】書いておぼえよう

発音したり，スペル練習したり，意味を調べたりしたらチェック欄の○をぬりなさい。
辞書を引いて単語の意味が多くある場合，英語の文脈から判断し，適切な意味を選びなさい。

英文	単語	品詞	意味を調べて書きなさい	発音しながらスペル練習しなさい	チェック
1)	(be) under way	熟語			○○○
1)	turbocharger	名詞			○○○
1)	abnormal ↔ normal	形容詞			○○○
1)	go off	熟語			○○○
1)	exhaust gas	名詞			○○○
1)	diesel generator engine	名詞			○○○
3)	overhaul	他動詞			○○○
3)	snap ring	名詞			○○○
3)	floating bearing	名詞			○○○
3)	discover / discovery	動 / 名			○○○
3)	fall off	熟語			○○○
4)	blower	名詞			○○○
4)	casing	名詞			○○○
5)	deformation / deform	名 / 動			○○○
5)	turbine rotor	名詞			○○○

理解を深めるキーワード「ターボチャージャーとフローティング軸受」

　ターボチャージャー（Turbo charger）とは，エンジンの排気ガスの持つ熱エネルギー（内部エネルギー）を利用してタービンを高速で回転させて圧縮空気をつくりエンジン内に送気することで吸気量を増やし，熱効率を向上させる装置である。今回の問題ではそのターボチャージャーで異音が発生し，アラームが鳴ったことが第1段落目に記載されている。

　第2段落目では，その原因調査のためにターボチャージャーを開放すると，フローティング軸受が外れていたことが確認された。フローティング軸受（Floating bearing）とは油膜により回転部分の摩擦を軽減する軸受で，金属同士が非接触となることから，細かな異物であれば軸・軸受を傷つけることがない。一方，軸受として広く用いられるボールベアリングタイプはボールと軸・軸受が接触する部分があり，異物が混入すると傷をつける可能性がある。

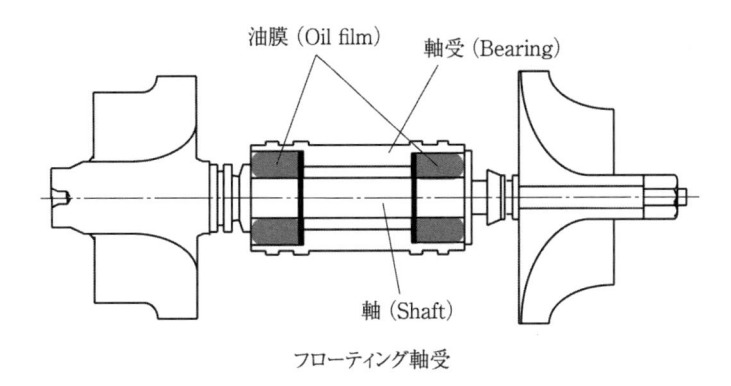

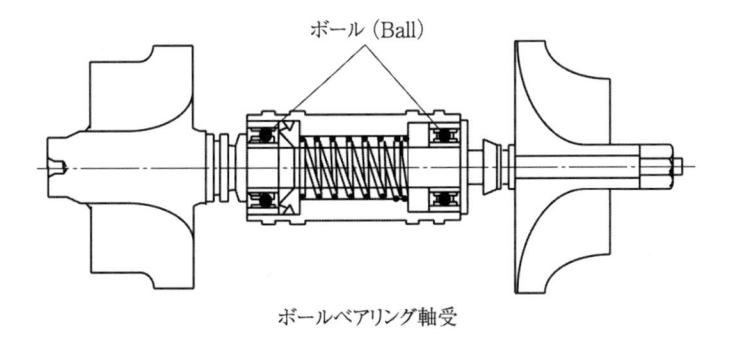

フローティング軸受

ボールベアリング軸受

Unit 25　給水処理

Step Check Box □ □ □

Step 1　下記の英文を読み，次の 3 つの活動をしなさい。終了後 Step Check Box □をぬりなさい。

(1) 動詞（述語動詞）を○で囲む。動詞の意味を考えて，動詞のあとに続く内容を推測する。
(2) 主語（主部）に下線を引く。
(3) スラッシュを入れる。スラッシュを入れる場所は主に下記の 3 つ。
- コンマ (,) やピリオド (.) でスラッシュ (/) を入れる。音読するときは，スラッシュで息つぎする。
- 接続詞の前にスラッシュを入れる。［接続詞＋主語＋動詞］は文としてまとまった意味を表す。
  ここでいう接続詞とは when, if, until, before, after など（and, but, or, so を含まない）。
- 原則として前置詞の前にスラッシュを入れる。（前置詞＋名詞）は，最小の意味のまとまりを表す。

1) Feed-water treatment deals with the various scale and corrosion causing salts and entrained gases by suitable chemical treatment.
2) This is achieved as follows:
2-1) By keeping the hardness salts in a suspension in the solution to prevent scale formation.
2-2) By stopping any suspended salts and impurities from sticking to the heat transfer surfaces.
2-3) By providing anti-foam protection to stop water carry-over.
2-4) By eliminating dissolved gases and providing some degree of alkalinity which will prevent corrosion.
3) The actual treatment involves adding various chemicals into the feed-water system and then testing samples of boiler water with a test kit.
4) The test kit is usually supplied by the treatment chemical manufacturer with simple instructions for its use.

出典：2 級海技士（機関）平成 22 年 7 月定期試験問題

Step 2　網掛けは主語，ゴシック体は動詞（述語動詞），＊語注・文法的留意事項，①②③は訳順番号

(1) 単語や熟語の説明を読みながら，部分訳を書きなさい。わからない単語の意味は調べなさい。
(2) 部分訳が終わったら，①②③の訳順番号を参考に，全文和訳を書きなさい。
　全文和訳が終わったら，Step Check Box □をぬりなさい。

1) Feed-water treatment① **deals with**⑥ the various scale and corrosion④/

* feed-water：〔名詞〕給水（水蒸気に変えるためにタンクなどからボイラーに供給される水）
  これが主語として使われているので，以下の英文では，それにかかわる情報が述べられていることがわかる。
  Feed-water treatment に関する専門知識があれば英文の理解ははやい。
* treatment：〔名詞〕処理　　　　　　　　　　　＊ deal with：〔動詞〕〜を扱う，〜に対処する
* various scale：さまざまなスケール　　　　　　　＊ corrosion：〔名詞〕腐食，さび
* scale：〔名詞〕湯あか（ボイラーの内側に生じる湯あか，加熱した鉄などの表面にできる酸化物の湯あか）

⑤⑥部分訳 _____

causing salts and entrained gases③/ by suitable chemical treatment.⑤/

* cause：〔他動詞〕〜を引き起こす原因となる。ここでは caused by salts and entrained gases という意味。
  ②③④部分訳：〔名詞〕塩，気水共発ガスが原因となるさまざまなスケールや腐食
* salt：〔名詞〕塩　　　　　　　　　　　　　　　＊ entrained gas：気水共発ガス
* entrain：〔他動詞〕（化学）飛沫同伴する，（蒸気などが小滴などを）運び去る，（液体が泡を吸収して）消す
* suitable：〔形容詞〕適切な　　　　　　　　　　＊ chemical：〔形容詞〕化学的な
* by：〔前置詞〕〜によって
* by suitable chemical treatment は⑥にかかる。「適当な化学的処理によって〜を扱う」となる。

全文和訳

2) This **is achieved**/ as follows:/
   　　①　　③　　　②

- * this：〔代名詞〕これは，feed-water treatment を指す。
- * is achieved：〔受動態〕行われる
- * as follows：〔熟語〕以下のように
- * achieve：〔他動詞〕～を果たす，成し遂げる，実行する
- * follow：〔自動詞〕～が続く

全文和訳　これ (feed-water treatment) は以下によって _____

2-1) By keeping the hardness salts/ in a suspension/ in the solution/ to prevent scale formation./
　　　　　　④　　　　　　　　　　③　　　　　　②　　　　　　　①

- * by：〔前置詞〕～することによって
- * by keeping ～：～を維持（一定に）することにより
- * in：〔前置詞〕～のなかにある，～以内，～に
  in は，基本的に「容器のなかに何かが入っている状態」を示す。
- * suspension：〔名詞〕浮遊物，懸濁液（液体中に顕微鏡で見える程度の粒子が分散しているもの，固体粒子が液体中に分散している状態）
- * suspend：〔他動詞〕浮遊させる，中断する
- * the hardness salts in a suspension：浮遊している塩分の硬度
- * solution：〔名詞〕溶液（液体をなす均一相の混合物）
- * scale：〔名詞〕湯あか，スケール，堆積物
- * formation：〔名詞〕
- * hardness：〔名詞〕硬度，硬さ。ここでは「塩分の硬度」。
- * solve：〔他動詞〕～を溶解する，解決する
- * prevent：〔他動詞〕～を防ぐ，阻止する
- * form：〔他動詞〕形成する，生じる

全文和訳 _____

2-2) By stopping any suspended salts and impurities/ from sticking to the heat transfer surfaces./
　　　　　　　③　　　　　　　　　　　　②　　　　　　　　　　　　①

- * stop A from B (Ving)：A を B することからストップする＝A が B しないようにする
- * suspended salt：懸濁した（浮遊した）塩分
- * impurities：（複数形で）不純物
- * stick to ～：〔自動詞〕～に付着する
- * pure：〔形容詞〕純粋な
- * transfer：〔名詞〕移動，輸送
- * purity：〔名詞〕純度
- * surface：〔名詞〕表面
- * heat transfer surface：伝熱面

全文和訳 _____

2-3) By providing anti-foam protection/ to stop water carry-over./
　　　　　③　　　　①　　　　　　　　　②

- * provide：〔他動詞〕～を用意（供給）する
- * protect：〔他動詞〕～を守る，防ぐ
- * carry-over：〔名詞〕ボイラ水や溶解している固形物がボイラ外へ持ち出される現象，キャリオーバ（汽水共発現象）
- * anti-foam protection：防護用の泡消し剤
- * protection：〔名詞〕～からの保護，防護
- * to stop ～：to V（不定詞）は「～するために」と訳す。to stop water carry-over は，providing にかかる。

全文和訳 _____

2-4) By eliminating dissolved gases/ and (by) providing some degree of alkalinity/ [which **will prevent** corrosion.]/
　　　　　①　　　　　　　　　　　　⑤　　　　③　　　　　　④　　　　　　　　　　②

- * eliminate：〔他動詞〕～を消去する，除去する
- * some degree of：ある度数の
- * prevent：〔他動詞〕～を防ぐ
- * which：〔関係代名詞〕関係代以下の英文は，関係詞の直前の名詞にかかる。which 以下が alkalinity を修飾するように訳す。
- * dissolved gas：溶存ガス
- * alkalinity：〔名詞〕アルカリ度（性）（pH 度のこと）
- * corrosion：〔名詞〕腐食，さび
- * provide：〔他動詞〕保持（用意）する

全文和訳 _____

3) The actual treatment **involves** adding various chemicals/ into the feed-water system/
   　　①　　　　　　　⑧　　④　　　　　　　　②　　　　　　　　　　　③

   * actual treatment：実際の処理
   * add ～ into …：〔他動詞〕…へ～を加える
   * adding：〔動名詞〕加えること
   * involve：〔他動詞〕～を必然的に含む，～を必要とする
   * chemicals：(通例，複数形で) 化学薬品
   * feed-water system：給水システム

   ②③④部分訳 _____

   and then testing samples of boiler water/ with a test kit./
   　　　　　⑦　　　　　⑤　　　　　　　　　⑥

   * test ～ with …：〔他動詞〕～を…で検査する
   * with：〔前置詞〕～で，使って，一緒に
   * sample：〔名詞〕試料，標本
   * test kit：テスト用のキット (一式)
   * testing：〔動名詞〕検査すること
   * with はさまざまなつながり，一緒という状態を示す。
   * boiler water：ボイラ水
   * 主語は the actual treatment，動詞は involves なので，ここでは実際の処理ではどうするのか書かれている。
     adding ～ and then testing ～：involve するものは「加えること」と「検査すること」である。

   ⑤⑥⑦部分訳 _____

   全文和訳 _____

4) The test kit **is** usually **supplied**/ by the treatment chemical manufacturer/ with simple instructions for its use./
   　　①　　　　　　　④　　　　　　　　　　②　　　　　　　　　　　　　　　③

   * is supplied with：〔受動態〕with 以下が提供される
   * treatment chemical manufacturer：処理剤薬品メーカー
   * with simple instructions for its use：使用のための簡単な取扱説明書を添えて
   * supply：供給する，提供する ＝ provide
     前置詞 with はさまざまな意味を持つが，基本的な意味は「～と一緒に」。ここでは be supplied with という熟語として働く。

   全文和訳 _____

Step 3   網掛け は主語，**ゴシック体**は動詞 (述語動詞)

(1) 音声を聞きながら，音読する。音読が終わったら，Step Check Box □をぬりなさい。
(2) 左から順に，スラッシュ毎の意味を考える。

Feed-water treatment **deals with** the various scale and corrosion/ causing salts and entrained gases/ by suitable chemical treatment./ This **is achieved**/ as follows:/
1. By keeping the hardness salts/ in a suspension/ in the solution/ to prevent scale formation./
2. By stopping any suspended salts and impurities/ from sticking to the heat transfer surfaces./
3. By providing anti-foam protection/ to stop water carry-over./
4. By eliminating dissolved gases/ and providing some degree of alkalinity/ which **will prevent** corrosion./

The actual treatment **involves** adding various chemicals/ into the feed-water system/ and then testing samples of boiler water/ with a test kit./ The test kit **is** usually **supplied**/ by the treatment chemical manufacturer/ with simple instructions for its use./

【文法のポイント】

◆ 前置詞＋名詞（＝前置詞句）は，前置詞の前にある動詞（あるいは名詞）にかかる

例1 Feed-water treatment **deals with** the various scale and corrosion (by suitable chemical treatment.)/
（前置詞 by ＋名詞 suitable chemical treatment）は前の動詞 deals with にかかる。
「給水処理は，（適切な化学的な処理によって）…に対処する」と訳す。

例2 The test kit **is** usually **supplied**/ by ～ manufacturer/ (with simple instructions for its use.)/
（前置詞 with ＋名詞 simple instructions for its use）は前の動詞 is usually supplied にかかる。
「検査キットは，使用のための簡単な取扱説明書を添えて，～によってふつうは提供される」と訳す。

◆ 名詞＋現在分詞（Ving）の語順ならば，現在分詞は「～する，している」と訳し，直前の名詞にかかる。

例3 the various scale and corrosion/ (causing salts and entrained gases)
causing は現在分詞形で，直前の名詞にかかる（修飾している）形容詞の働きをする。
「（塩分や気水共発ガスを引き起こす原因となる）さまざまなスケールやさび」と訳す。

【専門英語単語リスト】書いておぼえよう

発音したり，スペル練習したり，意味を調べたりしたらチェック欄の○をぬりなさい。
辞書を引いて単語の意味が多くある場合，英語の文脈から判断し，適切な意味を選びなさい。

英文	単語	品詞	意味を調べて書きなさい	発音しながらスペル練習しなさい	チェック
1)	treat / treatment	動／名			○○○
1)	cause	他動詞			○○○
2)	achieve	他動詞			○○○
2-1)	suspend / suspension	動／名			○○○
2-1)	solve / solution	動／名			○○○
2-1)	prevent / prevention	動／名			○○○
2-1)	form / formation	動／名			○○○
2-2)	transfer	名／動			○○○
2-3)	provide	他動詞			○○○
2-3)	protect / protection	動／名			○○○
2-4)	eliminate	他動詞			○○○
2-4)	corrosion	名詞			○○○
3)	feed-water system	熟語			○○○
4)	supply	他動詞			○○○
4)	manufacture	他動詞			○○○

Unit 26　ボイラ給水ポンプ

Step Check Box □ □ □

Step 1　下記の英文を読み，次の3つの活動をしなさい。終了後 Step Check Box □をぬりなさい。

(1) 動詞（述語動詞）を○で囲む。動詞の意味を考えて，動詞のあとに続く内容を推測する。
(2) 主語（主部）に下線を引く。
(3) スラッシュを入れる。スラッシュを入れる場所は主に下記の3つ。
  - コンマ（,）やピリオド（.）でスラッシュ（/）を入れる。音読するときは，スラッシュで息つぎする。
  - 接続詞の前にスラッシュを入れる。［接続詞＋主語＋動詞］は文としてまとまった意味を表す。
    ここでいう接続詞とは when, if, until, before, after など（and, but, or, so を含まない）。
  - 原則として前置詞の前にスラッシュを入れる。（前置詞＋名詞）は，最小の意味のまとまりを表す。

1) Question:
2) Explain why high-pressure centrifugal boiler feed pumps are usually multistaged.
3) Answer:
4) The feed pump must supply a high discharge pressure in order to force the water into the boiler.
5) To try and obtain this needed pressure from a single stage unit would require an exceptionally large impeller to be run at an excessive speed.
6) This would result in increased impeller friction and leakage losses due to the large difference in pressures between the suction and discharge sides of the pump.
7) The multistage pump consists of several impellers mounted on a single shaft.

出典：2級海技士（機関）平成22年10月定期試験問題

Step 2　網掛けは主語，ゴシック体は動詞（述語動詞），＊語注・文法的留意事項，①②③は訳順番号

(1) 単語や熟語の説明を読みながら，部分訳を書きなさい。わからない単語の意味は調べなさい。
(2) 部分訳が終わったら，①②③の訳順番号を参考に，全文和訳を書きなさい。
　　全文和訳が終わったら，Step Check Box □をぬりなさい。

1) Question:

　＊ question：〔名詞〕設問

2) **Explain**③/ [(the reason) why high-pressure centrifugal boiler feed pumps① **are** usually **multistaged**②.]/

　＊ explain：〔他動詞〕〜を説明する　　　　　　　　＊ 文頭が動詞で始まる文は命令文。「〜しなさい」と訳す。
　＊ why：〔関係副詞〕the reason が省略されている。「〜が…である理由」と訳す。
　　 why の後には，主語Sと動詞Vが必ずくる。
　＊ high-pressure：〔形容詞〕高圧の，高気圧の ↔ low-pressure：〔形容詞〕低圧の，低気圧の
　＊ centrifugal：〔形容詞〕遠心の
　＊ boiler feed pump：ボイラ給水ポンプ　　　　　　＊ multi-：多くの　　＊ multistaged：〔形容詞〕多段階の
　＊ multistage：〔形容詞〕（タービン，コンプレッサなどが）複数のロータ（回転子）を持つ
　　→ multi-stage centrifugal pump：多段うず巻ポンプ

全文和訳

3) Answer:

　＊ answer：〔名詞〕解答
　＊ 1）と 2）を訳すと，以下の英文で何が話題になっているのかわかる。
　＊「機関」に関する専門知識を十分に学んでおくことが必須である。
　＊ テキストで出合う動詞は基本語彙である。ビジネスの場面や英会話の場面でも応用できる語彙が多いので覚える。

4) The feed pump **must supply** a high discharge pressure/ in order to force the water/ into the boiler./

- * feed pump：〔名詞〕給水ポンプ
- * high discharge pressure：吐出高圧力
- * force ～ into …：〔他動詞〕～を…へ押し込む
- * into：〔前置詞〕～のなかへ
- * supply：〔他動詞〕～を提供する = provide
- * in order to 動詞の原形：〔熟語〕～するために
- * into は基本的に「～の内部へ」の方向を示す。

全文和訳

5) To try and obtain this needed pressure/ from a single stage unit **would require**

- * try：〔他動詞〕～を試みる
- * obtain：〔他動詞〕～を獲得する
- * To try and obtain this needed pressure：このように必要とされる圧力を獲得しようとするならば
- * need：〔他動詞〕～を必要とする = require, request
- * needed：〔過去分詞形〕必要とされる，pressure にかかる形容詞
- * single stage unit：単段のポンプ → single stage pump：単段のポンプ
  → single stage centrifugal pump：一段うず巻ポンプ
- * would require：「もし万一～ならば」という仮定の気持ちが主語に入っているので，助動詞 would が使われている。

①②③部分訳 _____

an exceptionally large impeller/ to be run/ at an excessive speed./

- * exceptionally：〔副詞〕例外的に，異常に
- * exceptional：〔形容詞〕例外的な，特別な
- * exception：〔名詞〕例外，除外
- * impeller：〔名詞〕インペラ，羽根車，（遠心式ポンプ，圧縮機，送風機の）翼車，（うず巻ポンプ，ジェットエンジンなどの）回転翼（羽根）
- * at an excessive speed：法外な速度で
- * excessive：〔形容詞〕法外な，極端な，度を越した
- * would require ～ to be run：～を作動する（～が動かされる）必要がある
- * run：〔他動詞〕（機械を）動かす
- * run-ran-run と変化し，to be run は受け身の形である。

全文和訳

6) This **would result in** increased impeller friction and leakage losses/

- * this would result in：this は 5) の内容を受けている。「5) のようにするならば，これは」という意味。
- * result in：結果として⑧⑨を生じる
- * increased：〔過去分詞〕増大した，直後の名詞 friction を修飾する。
- * friction：〔名詞〕摩擦力
- * leakage：〔名詞〕漏えい
- * loss：〔名詞〕損失
- * ⑧⑨⑩大意：impeller friction や leakage losses が増える結果となるだろう

due to the large difference in pressures/ between the suction and discharge sides/ of the pump./

- * due to：〔前置詞〕～のせい（が原因）で
- * large difference：大きい差
- * pressure：〔名詞〕圧力
- * large difference in pressure：大きな圧力差
- * between A and B：〔前置詞〕A と B の間で
- * suction side：吸込側
- * discharge side：吐出側

②③④⑤⑥⑦部分訳 _____

全文和訳

7) The multistage pump **consists of** several impellers/ mounted on a single shaft./
   　①　　　　　　　　　　　　　　⑤　　　　　　④　　　　③　　　　　　②

* multistage pump：多段式ポンプ　　　　　　　　　　* consist of：〜から成り立つ
* mounted：〔過去分詞〕「取り付けられた」と受け身のように訳す。
  過去分詞 mounted の直前に名詞 impellers が付いているので，過去分詞が名詞 impellers を修飾する働きをしている。
* on a single shaft：1つの軸に

全文和訳

Step 3　網掛けは主語，ゴシック体は動詞（述語動詞）

(1) 音声を聞きながら，音読する。音読が終わったら，Step Check Box □をぬりなさい。
(2) 左から順に，スラッシュ毎の意味を考える。

Question: **Explain**/ why high-pressure centrifugal boiler feed pumps **are** usually **multistaged**./
Answer: The feed pump **must supply** a high discharge pressure/ in order to force the water/ into the boiler./ To try and obtain this needed pressure/ from a single stage unit **would require** an exceptionally large impeller/ to be run/ at an excessive speed./ This **would result in** increased impeller friction and leakage losses/ due to the large difference in pressures/ between the suction and discharge sides/ of the pump./ The multistage pump **consists of** several impellers/ mounted on a single shaft./

【専門英語単語リスト】書いておぼえよう

発音したり，スペル練習したり，意味を調べたりしたらチェック欄の○をぬりなさい。
辞書を引いて単語の意味が多くある場合，英語の文脈から判断し，適切な意味を選びなさい。

英文	単語	品詞	意味を調べて書きなさい	発音しながらスペル練習しなさい	チェック
2)	explain / explanation	動／名			○○○
2)	high-pressure	名詞			○○○
4)	discharge	名詞			○○○
4)	in order to V	熟語			○○○
4)	force	他動詞			○○○
5)	obtain	他動詞			○○○
5)	impeller	名詞			○○○
5)	run	他動詞			○○○
6)	result in	他動詞			○○○
6)	friction	名詞			○○○
6)	leakage	名詞			○○○
6)	due to	前置詞			○○○
7)	consist of	他動詞			○○○
7)	mount	他動詞			○○○
7)	shaft	名詞			○○○

理解を深めるキーワード「給水ポンプ」

　タービンにて蒸気から大きな仕事を取り出す場合，その蒸気を高温高圧とする必要がある。船一隻を動かすだけの出力を生み出すためのタービンであれば，ボイラの圧力を20〜30MPaに維持しなければならない。このような高圧力場に圧縮水を供給するためには給水ポンプの吐出圧力もそれ以上の圧力が必要であり，ポンプ1段（Single stage）で高圧の吐出圧力を生成するには大きなインペラ（Large impeller）と高回転数（Excessive speed）が必要となる。このようなポンプでは摩擦損失（Friction loss）や漏えい損失（Leakage loss）が大きくなる。そこで，複数のインペラー（Impeller）を直列につなげる多段型（Multi stage）とすることにより，各段でみたときのインペラー入口圧力と吐出圧力の差を小さく抑制できる。その結果，損失も小さく抑えることができる。

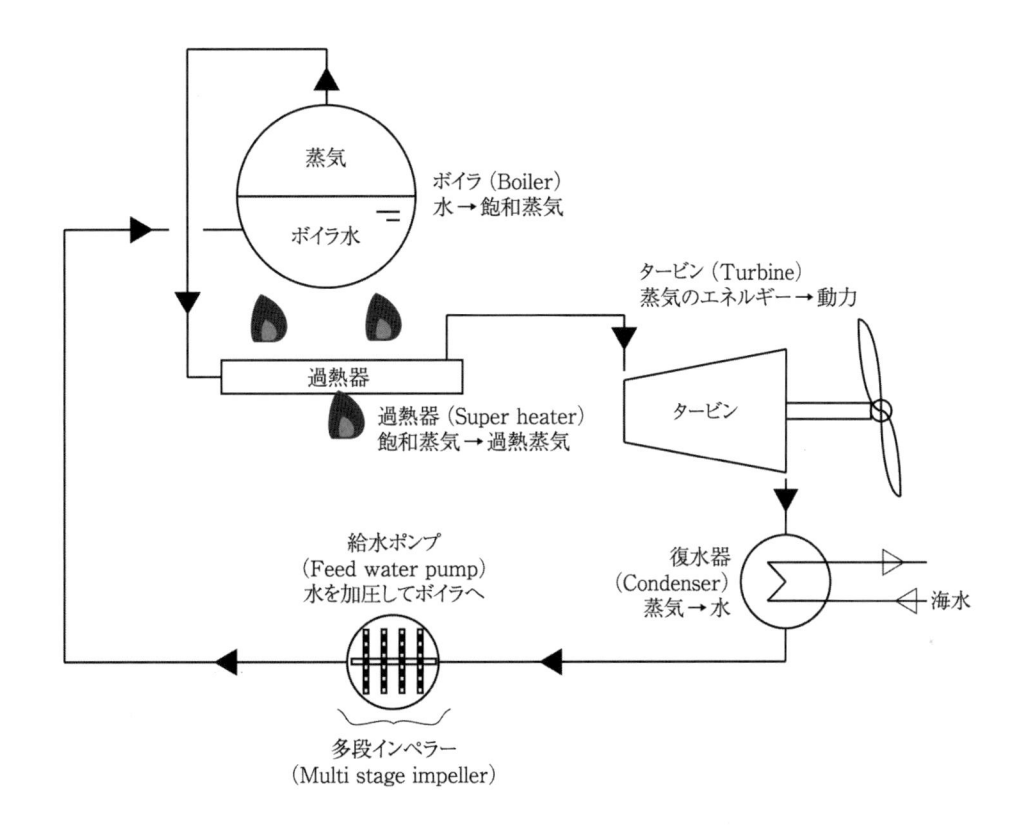

Unit 27　軸心調整

Step Check Box □ □ □

Step 1　下記の英文を読み，次の3つの活動をしなさい。終了後 Step Check Box □ をぬりなさい。

(1) 動詞（述語動詞）を○で囲む。動詞の意味を考えて，動詞のあとに続く内容を推測する。
(2) 主語（主部）に下線を引く。
(3) スラッシュを入れる。スラッシュを入れる場所は主に下記の3つ。
- コンマ (,) やピリオド (.) でスラッシュ (/) を入れる。音読するときは，スラッシュで息つぎする。
- 接続詞の前にスラッシュを入れる。[接続詞＋主語＋動詞]は文としてまとまった意味を表す。
  ここでいう接続詞とは when, if, until, before, after など (and, but, or, so を含まない)。
- 原則として前置詞の前にスラッシュを入れる。(前置詞＋名詞) は，最小の意味のまとまりを表す。

1) The intention of good alignments is to ensure that bearings are correctly loaded and that the shaft is not severely stressed.
2) Alignments can be checked with conventional methods, employing light and targets, laser or measurements from a taut wire.
3) There is, however, a continuity problem because the line of sight or taut wire cannot extend over the full length of an installed shaft.
4) There is no access to that part of the shaft within the stern tube and access is difficult in way of the propulsion machinery.
5) Results are also uncertain unless the vessel is in the same condition with regard to loading and hull temperatures as when the shaft system was installed or previously checked

出典：2級海技士（機関）平成23年10月定期試験問題

Step 2　網掛け は主語，**ゴシック体**は動詞（述語動詞），＊語注・文法的留意事項，①②③は訳順番号

(1) 単語や熟語の説明を読みながら，部分訳を書きなさい。わからない単語の意味は調べなさい。
(2) 部分訳が終わったら，①②③の訳順番号を参考に，全文和訳を書きなさい。
　　全文和訳が終わったら，Step Check Box □ をぬりなさい。

1) The intention of good alignments **is** to ensure/ [that bearings **are** correctly **loaded**/] and [that the shaft **is not** severely **stressed**./]
②　　　　　　①　　　　　　⑦　⑥　⑤　　　　　③
④

* intention：〔名詞〕目的　　　　　　　　　　　　＊ intend：〔他動詞〕～を意図する
* alignment：〔名詞〕軸心調整，調整，位置合わせ（2台以上の回転機械の回転中心線を一致させること）
* ensure：〔他動詞〕確認する　　　　　　　　　　＊ ensure すること（＝目的語）が that 以下に2つ書いてある。
* that：〔接続詞〕that（～ということを）以下には必ず主語 S と述語動詞 V がある。
* bearing：〔名詞〕ベアリング，軸受，方位，支点
* correctly：〔副詞〕正しく　　＊ correctly loaded：正しく負荷がかけられている　　＊ correct：〔形容詞〕正しい
* load：〔他動詞〕負荷する，荷重をかける
　（力学）荷重：エンジンや発動機などが一定条件のもとで打ち勝つ外部抵抗
* shaft：〔名詞〕シャフト，軸，車軸
* severely：〔副詞〕過酷に，厳しく　　　　　　　＊ severe：〔形容詞〕過酷な，厳しい
* stress：〔他動詞〕圧力を加える。is not stressed は受動態の否定。

全文和訳

2) Alignments **can be checked**/ with conventional methods,/ employing light and targets,/ laser or measurements from a taut wire./
①　　　　　　⑨　　　　　　⑧　　　　　　　　　⑦　②　③　④　　⑥
⑤

* can be checked：〔助動詞＋受動態〕検査される，調べられる
* with：〔前置詞〕～を使って

114

* 前置詞の後には必ず名詞がくる。［前置詞＋名詞］は1つのまとまりを指す。
* conventional：〔形容詞〕従来の，規格どおりの　　＊ conventional methods：用いる方法の具体例が4つ挙げてある。
* employ：〔他動詞〕〜を用いる → employing 〜：「light と target と，laser または measurement を用いて」となる。
* (employing light and targets, laser or measurements from a taut wire) は動詞 can be checked にかかる。
* target：〔名詞〕物標，目標物　　　　　　　　＊ measurement：〔名詞〕測量，計測，測定，測定値
* a taut wire：ピンと張ったワイヤ（鋼索）

全文和訳

3) There **is**,/ however,/ a continuity problem/ [because the line of sight or taut wire **cannot extend**/ over the full length of an installed shaft.]/
　　　③　　　　　①　　　②　　　　　　　④　　　　　　　　　⑤　　　　　　　　　⑧
　　　　　　　　　　　　　　　⑦　　　　　　　⑥

* there is：「あります」という意味。There は主語ではない。主語は a continuity problem で，動詞は is。
* a continuity problem：連続性の問題　　　　　　　＊ continue：〔自動詞〕続く
* because：〔接続詞〕〜というのは　　　　　　　　＊ because の後には主語Sと動詞Vが続く。
* sight：（航海者による方向・位置決定のための，測量機器や六分儀などによる）観測，天測など
* the line of sight：見通し線，照準ライン
* installed：（過去分詞なので受け身のように訳す）据え付けられた →⑥据え付けられたシャフト（軸）
* over the full length：全長以上に（全長を超えて）　＊ length〔名詞〕→ long〔形容詞〕
* extend：〔自動詞〕〜が広がる，及ぶ

全文和訳

4) There **is** no access/ to that part of the shaft/ within the stern tube/ and access **is** difficult/ in way of the propulsion machinery./
　　　④　③　　　　　②　　　　　　　　①　　　　　　　　　⑥
　　　　　　　　　　　⑤

* no access：名詞に否定語 no が付いているので，動詞を否定形にして「〜へ近づくことはできない」と訳す。
* that part of the shaft：軸の部分　　　　　　　　＊ access：〔名詞〕接近，利用
* within：〔前置詞〕〜のなかで，内部で　　　　　＊ the stern tube：船尾管
* in way of：〔熟語〕〜のそばに，〜と隣り合わせに
* access is difficult in way of 〜：〜のそばに接近することは難しい
* propulsion：〔名詞〕推進，前進，推進力　　　　＊ machinery：〔名詞〕機械類，機関
* the propulsion machinery：推進力となる機械（たとえばプロペラ）

全文和訳

5) Results **are** also uncertain/ [unless the vessel **is** in the same condition/ with regard to loading and hull temperatures/] as [when the shaft system **was installed**/ or (was) previously **checked**.]/
　　⑦　　　　　　　　⑧　　　　　　　①　　　　　　　⑥　　　　　　③
　　　　②　　　　　　　　　　　　　　　　　④　　　　　　　　　　⑤

* result：〔名詞〕結果　　　　　　　　　　　　　＊⑦⑧の大意：結果もまた正確ではない
* uncertain：〔形容詞〕不確かな ↔ certain：確かな，確実な
* unless：〔接続詞〕もし〜でないならば
　if not と同じ意味なので，日本語では動詞を否定形にして訳す。
　つまり，if the vessel is not in the same condition と同じ意味。
* in the same condition … as [when 〜]：（もし，船が）when 以下したときと同じ状態（でないならば）
* with regard to（名詞または動名詞 Ving）：〜に関しては
* loading and hull temperatures：積み荷と船体温度
* be installed：〔受動態〕取り付けられる，据え付けられる　　＊ previously：〔副詞〕以前に

①②③④⑤⑥部分訳

全文和訳

Step 3   網掛けは主語，**ゴシック体**は動詞（述語動詞）

(1) 音声を聞きながら，音読する。音読が終わったら，Step Check Box □をぬりなさい。
(2) 左から順に，スラッシュ毎の意味を考える。

The intention of good alignments **is** to ensure/ that bearings **are** correctly **loaded**/ and that the shaft **is not** severely **stressed**./ Alignments **can be checked**/ with conventional methods,/ employing light and targets,/ laser or measurements from a taut wire./ There **is**,/ however,/ a continuity problem/ because the line of sight or taut wire **cannot extend**/ over the full length of an installed shaft./ There **is** no access/ to that part of the shaft/ within the stern tube/ and access **is** difficult/ in way of the propulsion machinery./ Results **are** also uncertain/ unless the vessel **is** in the same condition/ with regard to loading and hull temperatures/ as when the shaft system **was installed**/ or previously **checked**./

【専門英語単語リスト】書いておぼえよう

発音したり，スペル練習したり，意味を調べたりしたらチェック欄の○をぬりなさい。
辞書を引いて単語の意味が多くある場合，英語の文脈から判断し，適切な意味を選びなさい。

英文	単語	品詞	意味を調べて書きなさい	発音しながらスペル練習しなさい	チェック
1)	intention / intend	名 / 動			○○○
1)	alignment	名詞			○○○
1)	ensure	他動詞			○○○
1)	load / loading	動 / 名			○○○
1)	stress	他動詞			○○○
2)	employ	他動詞			○○○
3)	continuity / continue	名 / 動			○○○
3)	extend	自動詞			○○○
4)	be no access to 名詞	熟語			○○○
4)	stern tube	名詞			○○○
4)	propulsion	名詞			○○○
4)	machinery / machine	名詞			○○○
5)	unless	接続詞			○○○
5)	with regard to 名詞	熟語			○○○
5)	install / installment	動 / 名			○○○

理解を深めるキーワード「軸心調整」

　一般に船舶の軸系（Shaft system）は主機関からつながるスラスト（Thrust）軸，中間（Intermediate）軸，プロペラ（Propeller）軸，スタンチューブ（Stern tube）およびプロペラからなり，これらの軸心（Axis）は船体中心線（1軸船の場合）と一致するように軸心調整（Alignment）がなされる。

　この精度が悪いと，軸を支える軸受（Bearing）面に偏った荷重（Load）が加わり，発熱やホワイトメタルの剥離を生じる。また，軸の振動や内部の繰り返し曲げ荷重による疲労破壊の原因となる。

　このようなことを防ぐため，船舶の建造時の軸系据付・加工には，軸系前部および後部に基準点を設け，これら2点をレーザー光線（Laser）や緊張（Taut）させたワイヤ（通常は細いピアノ線を用いる）などを用いて結び，これを基準としてそれぞれの軸位置を調整する。

　しかしながら，これらの調整値も船舶就航後では，積載貨物の状態，船体の温度などにより時々変化し，かつ，これを確認しようにも機器が据え付けられた後では障害物が多く，軸系全体にわたり見通すことは不可能である。

　したがって，現状では船舶建造時の軸心調整に当たっては，あらかじめ積荷などや温度などによる船体の歪を考慮した軸心調整を行うとともに，より精度を高めるために船殻工事による船体変形のなくなった状態で，しかも，普通は船体の温度がほぼ一様となる夜間（早朝3時～4時頃）を選んで実施される。

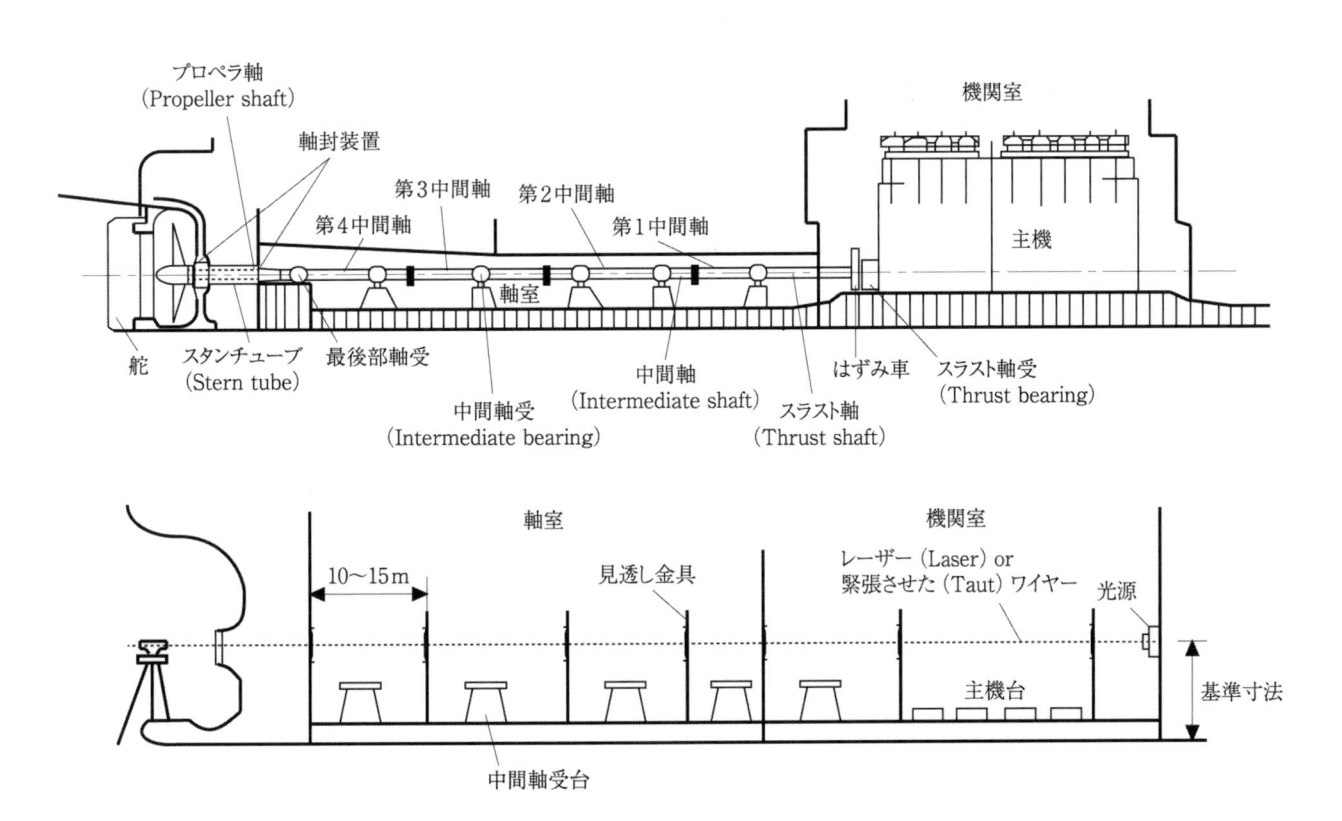

Unit 28　プロペラ軸

Step Check Box □ □ □

Step 1　下記の英文を読み，次の3つの活動をしなさい。終了後Step Check Box □をぬりなさい。

(1) 動詞（述語動詞）を○で囲む。動詞の意味を考えて，動詞のあとに続く内容を推測する。
(2) 主語（主部）に下線を引く。
(3) スラッシュを入れる。スラッシュを入れる場所は主に下記の3つ。
- コンマ(,)やピリオド(.)でスラッシュ(/)を入れる。音読するときは，スラッシュで息つぎする。
- 接続詞の前にスラッシュを入れる。［接続詞＋主語＋動詞］は文としてまとまった意味を表す。
  ここでいう接続詞とは when, if, until, before, after など（and, but, or, so を含まない）。
- 原則として前置詞の前にスラッシュを入れる。（前置詞＋名詞）は，最小の意味のまとまりを表す。

1) The transmission system on a ship transmits power from the engine to the propeller.
2) It is made up of shafts, bearings and finally the propeller itself.
3) The thrust from the propeller is transferred to the ship through the transmission system.
4) The different items in the system include the thrust block, intermediate bearings and the stern tube bearings.
5) A sealing arrangement is provided at either end of the tail shaft with the propeller and cone completing the arrangement.

出典：2級海技士（機関）平成23年4月定期試験問題

Step 2　網掛けは主語，**ゴシック体**は動詞（述語動詞），＊語注・文法的留意事項，①②③は訳順番号

(1) 単語や熟語の説明を読み，わからない単語の意味は調べなさい。
(2) ①②③の訳順番号を参考に，全文和訳を書きなさい。
　全文和訳が終わったら，Step Check Box □をぬりなさい。

1) The transmission system/ on a ship **transmits** power/ from the engine/ to the propeller./
   　②　　　　　　　　　①　　　⑥　　　　③　　　　　④　　　　　⑤

- ＊ transmission：〔名詞〕（動力の）伝達
- ＊ system：〔名詞〕装置
- ＊ transmit A to B：〔他動詞〕A（熱，電気など）をBへ伝える
- ＊ propeller：〔名詞〕プロペラ
- ＊ power：動力（単位時間になされる機械的仕事量）。単位はワット(W)，馬力(PS)など。
  　　　　　電力（単位時間に機器や装置に発生または消費される電気エネルギー）。単位はワット(W)。
- ＊ この文の主語Sは transmission system，動詞Vが transmit なので，ここでは，動力の伝動の仕組みが話題なることを予想し，専門科目で学んでいることを思い出す。主語と動詞の後には，取り上げたトピックを詳しく説明する情報が追加されていく。

全文和訳

2) It **is made up of** shafts, bearings and finally the propeller itself./
   ①　　⑤　　　　　②　　　③　　　　　　　④

- ＊ it：ここでは前出の主語 the transmission system（動力の伝達装置）を示す。
- ＊ be made up of：〔受動態〕〔熟語〕から成り立っている
- ＊ shaft：〔名詞〕軸
- ＊ bearing：〔名詞〕軸受
- ＊ finally：〔副詞〕最後に
- ＊ itself：〔再帰代名詞〕そのもの，それ自身

全文和訳

118

3) The thrust/ from the propeller **is transferred to** the ship/ through the transmission system./

- ＊ thrust：〔名詞〕推進力，推力
- ＊ be transferred to ～：〔受動態〕～へ伝えられる
- ＊ from：〔前置詞〕～から
  from は「起点から出発して離れていく」動きを示す。
  be different from，prevent A from B などのように，分離・区別を示す場合に使われる。
- ＊ to：〔前置詞〕～へ
  to は「ある行為がどこに到達するのか」という到達点を指し示す。
- ＊ transfer：〔他動詞〕（熱，電気などを）伝える
- ＊ through：〔前置詞〕～を通り抜けて

全文和訳

4) The different items/ in the system **include** the thrust block, intermediate bearings and the stern tube bearings./

- ＊ different：〔形容詞〕別の，異なる
- ＊ include：〔他動詞〕～を含む
- ＊ intermediate：〔形容詞〕中間の
- ＊ thrust block：スラスト軸受，推力軸受（荷重が軸に沿う推力である場合の軸受。軸受の荷重を受ける面は，軸に垂直な平面に対し，円形または輪形の形象をなす）
- ＊ item：〔名詞〕部品，品目，項目
- ＊ system：〔名詞〕装置
- ＊ stern tube bearings：船尾管軸受

全文和訳

5) A sealing arrangement **is provided**/ at either end of the tail shaft/ **with** the propeller and cone/ completing the arrangement./

- ＊ sealing arrangement：軸封装置
- ＊ be provided with ～：〔受動態〕～が備え付けられている
- ＊ at either end：の両端に，の両方に
- ＊ propeller：〔名詞〕プロペラ
- ＊ complete：〔他動詞〕～を完成させる，仕上げる
- ＊ arrange：〔他動詞〕
- ＊ tail shaft：後尾の軸＝プロペラ軸
- ＊ cone：〔名詞〕軸コーンパート（テーパ部）
- ＊ arrangement：〔名詞〕設備，装置
- ＊ completing the arrangement：（シール）装置が完璧になるように（仕上げる）
  現在分詞 completing は名詞 the propeller and cone を修飾している。Ving は「～している」と訳す。
  現在分詞 Ving や過去分詞 Ved は動詞の意味を持ちながら，Ving「～している」や，Ved「～される」のように訳し，形容詞のように，直前の名詞を修飾する働きをしている。

全文和訳

Step 3 網掛けは主語，**ゴシック体**は動詞（述語動詞）

(1) 音声を聞きながら，音読する。音読が終わったら，Step Check Box □をぬりなさい。
(2) 左から順に，スラッシュ毎の意味を考える。

The transmission system/ on a ship **transmits** power/ from the engine/ to the propeller./ It **is made up of** shafts, bearings and finally the propeller itself./ The thrust/ from the propeller **is transferred** to the ship/ through the transmission system./
The different items/ in the system **include** the thrust block, intermediate bearings and the stern tube bearings./ A sealing arrangement **is provided**/ at either end of the tail shaft/ **with** the propeller and cone/ completing the arrangement./

【専門英語単語リスト】書いておぼえよう

発音したり，スペル練習したり，意味を調べたりしたらチェック欄の○をぬりなさい。
辞書を引いて単語の意味が多くある場合，英語の文脈から判断し，適切な意味を選びなさい。

英文	単語	品詞	意味を調べて書きなさい	発音しながらスペル練習しなさい	チェック
1)	transmission system	連語			○○○
1)	transmit / transmission	動 / 名			○○○
1)	power	名詞			○○○
1)	engine	名詞			○○○
1)	propeller	名詞			○○○
2)	be made up of〜	受動態			○○○
2)	bearing	名詞			○○○
3)	transfer	他動詞			○○○
3)	thrust	名詞			○○○
4)	include / including	他 / 前			○○○
4)	intermediate	形容詞			○○○
5)	provide	他動詞			○○○
5)	tail shaft	連語			○○○
5)	complete	他動詞			○○○
5)	arrangement / arrange	名 / 動			○○○

コラム2

助動詞

この教科書の英文の単語の登場回数のランキング（6ページ参照）のなかで，ゴシック体の単語は助動詞という種類の単語です。ほとんどの助動詞をみなさんはもう知っていると思いますが，助動詞の後には原形の動詞が続いて，(助動詞＋原形動詞)で述語動詞になることは覚えておきましょう。

練習のために次の英文を読んでみましょう。

Appropriate precautions shall be taken during loading and transport of heavy cargoes or cargoes with abnormal physical dimensions.

shall が助動詞であることと，(助動詞＋原形動詞) で述語動詞になることを覚えていれば，単語の意味がよくわからなくても，この英文の述語動詞は shall be taken の部分だとわかります。

また，(前置詞＋名詞) の部分をとりあえず無視して，残った部分から主語を探すと，Appropriate precautions がこの英文の主語だということがわかります。

Appropriate precautions **shall be taken** during loading and transport of heavy cargoes or cargoes with abnormal physical dimensions.

「助動詞」に関しては他に次の2点に注意しておきましょう。

1. shall は規則や契約に関係する文章で「〜するものとする」という意味で用いられます。
   したがって，上の英文は「Appropriate precautions が take されるものとする」といった感じになります。
   shall は航海の英語ではよく目にする助動詞なので，航海の人は必ず覚えておきましょう。

2. would, could, might などの助動詞の過去形は，過去の意味で使われることは少ないので注意しましょう。

Unit 29 冷凍機

Step Check Box ☐ ☐ ☐

Step 1　下記の英文を読み，次の3つの活動をしなさい。終了後 Step Check Box ☐ をぬりなさい。

(1) 動詞（述語動詞）を○で囲む。動詞の意味を考えて，動詞のあとに続く内容を推測する。
(2) 主語（主部）に下線を引く。
(3) スラッシュを入れる。スラッシュを入れる場所は主に下記の3つ。
- コンマ（,）やピリオド（.）でスラッシュ（/）を入れる。音読するときは，スラッシュで息つぎする。
- 接続詞の前にスラッシュを入れる。［接続詞＋主語＋動詞］は文としてまとまった意味を表す。
  ここでいう接続詞とは when, if, until, before, after など（and, but, or, so を含まない）。
- 原則として前置詞の前にスラッシュを入れる。（前置詞＋名詞）は，最小の意味のまとまりを表す。

1) The compressor is started and stopped by the LP (low pressure) controller in response to changes of pressure in the compressor suction.
2) There is also an HP (high pressure) cut-out with a hand re-set which operates to shut down the compressor in the event of excessively high discharge pressure.
3) The compressor can supply a number of cold compartments through thermostatically controlled solenoids.
4) Thus as each room temperature is brought down, its solenoid will close off the liquid refrigerant to that space.
5) When all compartment solenoids are shut, the pressure drop in the compressor suction will cause the compressor to be stopped through the LP controller.
6) A subsequent rise of compartment temperature will cause the solenoids to be re-opened by the room thermostats.
7) A pressure rise in the compressor suction acts through the LP controller to restart the compressor.

出典：2級海技士（機関）平成23年7月定期試験問題

Step 2　網掛け は主語，**ゴシック体**は動詞（述語動詞），＊語注・文法的留意事項，①②③は訳順番号

(1) 単語や熟語の説明を読みながら，部分訳を書きなさい。わからない単語の意味は調べなさい。
(2) 部分訳が終わったら，①②③の訳順番号を参考に，全文和訳を書きなさい。
　全文和訳が終わったら，Step Check Box ☐ をぬりなさい。

1) ① The compressor ⑥ **is started**/ and (**is**) ⑦ **stopped**/ by the LP (low pressure) controller/ ⑤ in response to ④ ③ changes of pressure/ ② in the compressor suction./

* compressor：〔名詞〕圧縮機
* LP (low pressure)：〔形容詞・名詞〕低圧（の）
* pressure：〔名詞〕圧縮，圧力，気圧，プレッシャー
* controller：〔名詞〕制御器，制御装置
* in response to ～：〔熟語〕～に応じて
* response：〔名詞〕～からの応答
* suction：〔名詞〕吸込み，吸入，サクション
* ここでは，compressor の開閉が取り上げられている。
* press：〔他動詞〕～を押しつける，圧縮する
* low pressure controller：低圧スイッチ
* respond to ～：〔自動詞〕～に応答する
* changes of pressure：圧力の変化
* suck：〔他動詞〕

全文和訳

2) ⑦ There **is** also an ⑥ HP (high pressure) cut-out/ ⑤ with a hand re-set/ ④ [which **operates** ③ to shut down the compressor/

* there is：この構文では there は飾りの主語，is が動詞で「～があります」と訳す。There is の後に「何がある」のか述べる。
* HP (high pressure)：〔形容詞・名詞〕高圧（の），高気圧（の）
* HP cut-out：高圧スイッチ
* hand re-set：手動復帰
* with：〔前置詞〕～のついている，～を使った，～で
* 関係代名詞 which 以下の英文は，前の名詞（ここでは hand re-set）にかかる（を修飾する）。
* operate：〔自動詞〕作動する
* cut-out：〔名詞〕遮断装置，排気弁，安全開閉器
* 前置詞にはさまざまな意味がある。
* to shut down：〔不定詞〕～を止めるために

③④⑤⑥部分訳 _____

　　　　　　　　②　　　　　　　　　　　　　①
in the event of excessively high discharge pressure.]/

* in the event of ～：〔熟語〕～の場合には
* excessively：〔副詞〕過度に
* excessively high discharge pressure：過度に高い吐出圧
* discharge：〔名詞〕吐出，〔他動詞〕排出する

全文和訳 _____

　　　①　　　　　　　　④　　　　　　　　　③　　　　　　　　　　　　　　②
3) The compressor **can supply** a number of cold compartments/ through thermostatically controlled solenoids./

* supply：〔他動詞〕～を供給する，提供する，補充する　　* a number of ～：〔熟語〕数多くの
* cold compartment：〔名詞〕冷凍庫，冷たい区画（仕切り）　* through：〔前置詞〕～によって（手段）
* thermostatically (controlled)：〔副詞〕サーモスタットで（制御された）
* thermo-：熱の　　　　　　　　　　　　　　　　　* thermostat：〔名詞〕温度自動調整器
* solenoid：〔名詞〕電磁弁（電流が流れると金属棒が動く仕掛けの制御スイッチで，自動制御システムなどで使われる）

全文和訳 _____

　　①　　④　　　　　②　　　　　　　③　　　　　　　⑤　　　⑧　　　　　　　⑦
4) Thus/ [as each **room temperature is brought down**,]/ its solenoid **will close off** the liquid refrigerant/
　　⑥
to that space./

* thus：〔副詞〕このようにして　　　　　　　　　　* is brought down：〔受動態〕（熱，血圧などが）下げられる
* as：〔接続詞〕～のとき
　asの意味は「～のとき」「～なので」「～しながら」など多種なので，文脈に合うよう訳す。
* each room：3) の cold compartments（冷凍庫）のこと。　* each room temperature：庫内の気温
* solenoid：〔名詞〕電磁弁　　　　　　　　　　　　* close off：〔他動詞〕～の流れをせき止める，遮断する
* liquid：〔名詞〕液体，流動体　　　　　　　　　　* solid：〔名詞〕固体　　　* gas：〔名詞〕気体
* liquid refrigerant：液体冷媒　　　　　　　　　　* refrigerant：〔名詞〕（アンモニアなどの）冷媒
* to that space：冷凍庫内への　　　　　　　　　　* space とは，ここでは cold compartments のこと。

②③④部分訳 _____

⑤⑦⑧部分訳 _____

全文和訳 _____

　　③　　　　　①　　　　　　　　②　　　　　　⑤　　　　　　　④　　　　　　⑧
5) [When all compartment solenoids **are shut**]/ the pressure drop/ in the compressor suction/ **will cause**
　　⑦　　　　　　　　　　　⑥
the compressor to be stopped/ through the LP controller./

* compartment solenoid：庫内の電磁弁　　　　　　* pressure drop：圧力降下
* compressor suction：圧縮機の吸込み　　　　　　* in：〔前置詞〕～のなかで
* cause：〔他動詞〕～の原因となる，～を引き起こす　* through：〔前置詞〕～によって（手段）
* to be stopped：〔不定詞の受動態〕止められる（＝止まる）* LP controller：低圧スイッチ
* the compressor to be stopped：圧縮機が止まる

④⑤⑧部分訳 _____

全文和訳 _____

123

6) A subsequent rise/ of compartment temperature **will cause** the solenoids to be re-opened/ by the room thermostats./

* a subsequent rise：続けて上昇することは
* compartment temperature：庫内の温度
* cause the solenoids to be re-opened：電磁弁が再び開く原因となる
* by the room thermostat：庫内のサーモスタットによって

全文和訳

7) A pressure rise/ in the compressor suction **acts**/ through the LP controller/ to restart the compressor./

* a pressure rise：圧力上昇
* compressor suction：圧縮機の吸込み
* A pressure rise in the compressor suction acts：圧縮機の吸込み圧力の上昇
* act：〔自動詞〕（機械，車などが正常に）作用する，動く，働く
* restart：〔他動詞〕を再開させる
  不定詞は「～することは」「～するための」「～するために」と3つの訳し方がある。
  ここでは，to restart は動詞 act にかかるので「～するために」「～するように」と訳す。
* through：〔前置詞〕～によって（手段）

全文和訳

Step 3 網掛け は主語，ゴシック体は動詞（述語動詞）

(1) 音声を聞きながら，音読する。音読が終わったら，Step Check Box □をぬりなさい。
(2) 左から順に，スラッシュ毎の意味を考える。

The compressor **is started**/ and **stopped**/ by the LP (low pressure) controller/ in response to changes of pressure/ in the compressor suction./ There **is** also an HP (high pressure) cut-out/ with a hand re-set/ which **operates** to shut down the compressor/ in the event of excessively high discharge pressure./ The compressor **can supply** a number of cold compartments/ through thermostatically controlled solenoids./ Thus/ as each room temperature **is brought down**,/ its solenoid **will close off** the liquid refrigerant/ to that space./ When all compartment solenoids **are shut**,/ the pressure drop/ in the compressor suction/ **will cause** the compressor to be stopped/ through the LP controller./ A subsequent rise/ of compartment temperature **will cause** the solenoids to be re-opened/ by the room thermostats./ A pressure rise/ in the compressor suction **acts**/ through the LP controller/ to restart the compressor./

【専門英語単語リスト】書いておぼえよう

発音したり，スペル練習したり，意味を調べたりしたらチェック欄の○をぬりなさい。
辞書を引いて単語の意味が多くある場合，英語の文脈から判断し，適切な意味を選びなさい。

英文	単語	品詞	意味を調べて書きなさい	発音しながらスペル練習しなさい	チェック
1)	compressor	名詞			○○○
1)	pressure	名詞			○○○
1)	suction	名詞			○○○
2)	cut-out	名詞			○○○
2)	operate / operation	動／名			○○○
2)	in the event of ～	熟語			○○○

英文	単語	品詞	意味を調べて書きなさい	発音しながらスペル練習しなさい	チェック
3)	supply = provide	他動詞			○○○
3)	a number of ~	熟語			○○○
4)	temperature	名詞			○○○
4)	liquid	名詞			○○○
4)	refrigerant	名詞			○○○
5)	solenoid	名詞			○○○
5)	cause	他動詞			○○○
5)	re-open	他動詞			○○○
5)	act	自動詞			○○○

理解を深めるキーワード「エアコン」

　エアコンの基本サイクルについて述べる。圧縮機（Compressor）で冷媒（Refrigerant）を高温・高圧の気体の状態にする。その気体を凝縮器（Condenser）と呼ばれる外気との熱交換器へ送ると，気体は凝縮し，配管内の冷媒は液化して温度が下がる。この冷媒は膨張弁により流量が調整され，圧力が下がり低圧の液体となる。低圧の冷媒は飽和温度（沸騰する温度）が下がるため，蒸発器（Evaporator）と呼ばれる周辺空気との熱交換器を通過することで蒸発（気化）する。この蒸発時に周辺空気の熱を奪って相変化が起こるため，周囲空間の温度が低下する。この低下させた外気をファンで機外へ送り出すことで冷風が生成され，冷却を必要とする場所へ送る（Cold compartment）。配管内を循環する冷媒は蒸発機通過後に圧縮機へ戻り，再び加圧され，高温状態の気体として凝縮器へ導かれる。

　冷媒を連続的に循環させれば冷風が絶えず生成されることになるため，冷却したい場所の温度を検出するサーモスタットにより蒸発機に送られる冷媒を遮断する電磁弁（Thermostatically controlled solenoids）が配管内に設置されている。つまり冷却したい場所，たとえば室内の温度が設定温度よりも低くなると電磁弁を遮断して冷媒の循環をストップさせる。冷媒の循環が止まっても圧縮機が動き続ければ圧縮機手前の冷媒圧力は下がり続けるため，この吸込み圧力を検出するための圧力センサー（Low pressure controller）が機能すると圧縮機が停止する。これにより冷風が供給されなくなると室内の温度は上昇する。ある温度まで上昇すると再びサーモスタットにより電磁弁が開き，電磁弁手前の高圧状態の冷媒が膨張弁を通過して圧縮機吸込み側に到達する。その結果，再び圧縮機が動き出し，冷風が生成される。

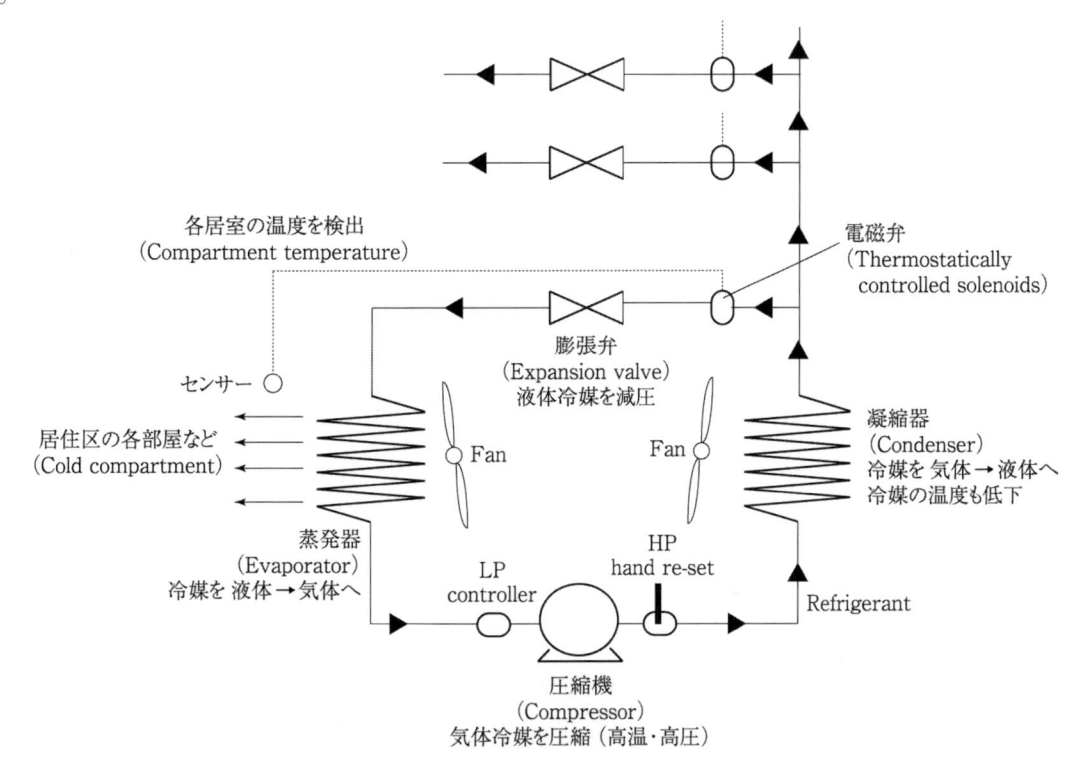

Unit 30　LO 清浄系統

Step Check Box □ □ □

Step 1　下記の英文を読み，次の3つの活動をしなさい。終了後 Step Check Box □ をぬりなさい。

(1) 動詞（述語動詞）を○で囲む。動詞の意味を考えて，動詞のあとに続く内容を推測する。
(2) 主語（主部）に下線を引く。
(3) スラッシュを入れる。スラッシュを入れる場所は主に下記の3つ。
- コンマ (,) やピリオド (.) でスラッシュ (/) を入れる。音読するときは，スラッシュで息つぎする。
- 接続詞の前にスラッシュを入れる。［接続詞＋主語＋動詞］は文としてまとまった意味を表す。
  ここでいう接続詞とは when, if, until, before, after など (and, but, or, so を含まない)。
- 原則として前置詞の前にスラッシュを入れる。（前置詞＋名詞）は，最小の意味のまとまりを表す。

1)　The treatment process of the system oil is classified into the following two methods.
2)　It is necessary to properly implement the following two treatment processes in various situations.
3)　Bypass Purifying
3-1) Set the oil-flow rate for the purifier to approximately 1,700 liters/h/10,000 kw. (1,300 liters/h/10,000 ps) and the oil-flow temperature to between 80 to 85 ℃.
4)　Entire Quantity Purifying
5)　Treat all the system oil during a docking period when the main engine is stopped for a prolonged period of time.
5-1) Transfer all the system oil to the LO settling tank and purify it by circulating it through the purifier.
5-2) In addition, clean the lubrication oil sump tank.
5-3) Set the heating temperature in the tank to approximately 80 ℃.

出典：2級海技士（機関）平成23年4月定期試験問題

Step 2　網掛け は主語，**ゴシック体**は動詞（述語動詞），＊語注・文法的留意事項，①②③は訳順番号

(1) 単語や熟語の説明を読みながら，部分訳を書きなさい。わからない単語の意味を調べなさい。
(2) 部分訳が終わったら，①②③の訳順番号を参考に，全文和訳を書きなさい。
　全文和訳が終わったら，Step Check Box □ をぬりなさい。

1)　① The treatment process of the system oil ③ **is classified**/ ② into the following two methods./

* treatment：〔名詞〕
* treat〔動詞〕処理する
* 基本となる単語の意味を覚えておくと，語形が似たような単語が出てきたときに，意味を類推しやすい。
* この文の主語は the treatment process of the system oil（システムオイルの処理手順）で，動詞が is classified（分類される）。システムオイルの処理手順を専門知識として知っていると，英文を理解しやすい。
* process：〔名詞〕手順，方法，プロセス
* system oil：システムオイル，循環油
* be classified：〔受動態〕～に分類される
* into：〔前置詞〕(= in + to)「内部へ」「～へ」という方向を示す。
* following：〔形容詞〕次に続く，以下の = next
* method：〔名詞〕方法

全文和訳

2)　③ It **is** necessary/ ② to properly implement the following two treatment processes/ ① in various situations./

* It は仮主語の働きをしている。to 以下が真主語である。It を「それは」と訳さない。
  「It …to V ～」という形式をとる。「to V ～」は「～することは」と訳す。
* properly：〔副詞〕正しく
* proper：〔形容詞〕正しい
* implement：〔他動詞〕～を実行する
* two treatment processes：ここでは，3) bypass purifying と 4) entire quantity purifying の処理手順である。
  この2つの処理手順がトピックだとわかると，書かれている内容が推測しやすい。
* situation：〔名詞〕状況
  situation の類語：condition（状態，状況），circumstance（事情，環境，状況），state（状態，事情）

126

3) After removal/ they **should be kept**/ in a clean place./

- * removal：〔名詞〕除去
- * remove：〔他動詞〕～を取り除く
- * they：〔代名詞〕ここでは前出の複数形名詞 mechanical seals を指す。
- * should be kept：〔助動詞＋受動態〕保管されるべきである
- * in a clean place：これは動詞 should be kept にかかる。

全文和訳

4) The two seal faces **should be** carefully **cleaned**/ by washing them/ in the same clean fluid/ [that **is being pumped**.]/

- * face：〔名詞〕表面，外面，表側
- * seal face：密封した表面
- * by (washing them)：〔前置詞〕（手段や方法を表す）～によって，～で
- * in the same clean fluid [that is being pumped]：関係代名詞 that 以下の文は fluid にかかる。
- * is being pumped：現在進行形の受動態で，「いま～されている」という意味。
- * ②③の大意：いまポンプでくみ上げられている同じきれいな液体のなかで

全文和訳

5) **Avoid** wiping the seal faces/ [as any scratches **could cause** the seal to leak.]/

- * avoid Ving（動名詞）：〔他動詞〕～するのを避ける
- * wipe：〔他動詞〕～を拭く
- * scratch：〔名詞〕ひっかきキズ
- * could：〔助動詞〕can の過去形だが，ここでは現在（または未来）に対する可能性や推量を表す。
- * seal face：密封した表面
- * as：〔接続詞〕～なので（主語 S と動詞 V が後に続く）
- * cause A to B：〔他動詞〕A が B を引き起こす
- * as any scratches could cause the seal to leak：どんなキズも密封の漏れを引き起こす可能性があるので

全文和訳

6) Conventional gland packing **should be replaced**/ at the time of pump overhaul./

- * conventional：〔形容詞〕従来の
- * overhaul：〔名詞〕分解修理
- * should be replaced：〔助動詞＋受動態〕取り換えられるべきである
- * at the time of ～：〔熟語〕～のときに

全文和訳

7) **Ensure** [(that) the gland stuffing box **is** clean/ and all old packing **has been removed**.]/

- * ensure：命令文なので，主語はない。
   大意：[that ＋主語 S ＋動詞 V 以下]を確認しなさい
- * stuffing box：パッキン箱，詰め箱（packing box）
- * gland：パッキン押さえ
- * has been removed：〔現在完了形の受動態〕これまでに（すでに）取り除かれてしまっている
- * remove：〔他動詞〕～を取り除く

全文和訳

8) **Wrap**③ a turn of packing②/ around the shaft①/ and **make**⑤ it slightly larger/ than the shaft circumference④./

- * wrap a turn of packing around the shaft：シャフトの周りをパッキンで一回り包む
  wrap：〔他動詞〕～を包む，くるむ
- * 7)～11)は命令文である。
- * packing：〔名詞〕（締め付けをよくしたり，気体・液体の漏れを防ぐ）パッキン
- * make it slightly larger than ～：make は使役動詞なので「それ(packing)を少し～より大きくさせる」と訳す
- * shaft circumference：シャフトの周囲，外周，円周

全文和訳

9) **Cut** the lengths of packing③/ with a sharp knife②/ to give a clean butt joint①./

- * length：〔名詞〕端から端までの長さ
  the lengths と複数形になっているので，「両端」の意味。
- * with (a sharp knife)：〔前置詞〕～を使って
- * to give a clean butt joint：きれいな突き合わせ継手となるように
  to give 以下は動詞 cut にかかる。
- * butt：〔名詞〕板の接合部，（柄などになっている）太い方の端
- * butt joint：突き付け，突き合わせ継手。芋目地（厚板と厚板の，端と端／側面と側面）を突き合わせて接合する方法。

全文和訳

10) **Insert**⑥ each turn of packing④/ **into**⑤ the stuffing box③/ offsetting the butt joints①/ by 120°②./

- * insert A into B：〔他動詞〕AをBに差し込む
- * stuffing box：パッキン箱，詰め箱(packing box)
- * each turn：それぞれのつなぎ部分
- * offset：〔名詞〕（管，でこ，さおなどを，障害物を避け，他の部分との連結を考慮して）少しずらすこと
  〔他動詞〕（管などを）軸（中心線）をはずして置く
- * offsetting the butt joins by 120°：offsetting は現在分詞形で，「～しながら」と訳し，動詞 insert にかかる。「突き合わせ継手を 120 度ずらして」となる。

全文和訳

11) **Put in**⑤ sufficient turns④/ [so that the gland ring nuts ① **will fit** fully③/ **on** the studs②.]/

- * sufficient：〔形容詞〕十分な
- * put in sufficient turns：十分なつなぎ部分をとりなさい
- * gland ring nut：パッキン押さえ用のリングナット（円形状のナット）
- * nut：親（留め）ねじ（nut とはまりあうのが bolt）
- * fit fully on ～：～にしっかりと合う，はまる
- * stud：植え込みボルト，スタッド（栓，ネジ，ピンなどの小突起）
- * so that 主語 S ＋動詞 V：〔接続詞〕～するように

全文和訳

Step 3 網掛けは主語，**ゴシック体**は動詞（述語動詞）

(1) 音声を聞きながら，音読する。音読が終わったら，Step Check Box □をぬりなさい。
(2) 左から順に，スラッシュ毎の意味を考える。

Pump Glands

Many pumps now **have** mechanical seals/ which **are** very expensive/ and **should be treated**/ with great care./ After removal/ they **should be kept**/ in a clean place./ The two seal faces **should be** carefully **cleaned**/ by washing them/ in the same clean fluid/ that **is being pumped**./ **Avoid** wiping the seal faces/ as any scratches **could cause** the seal to leak./ Conventional gland packing **should be replaced**/ at the time of pump overhaul./

- **Ensure** the gland stuffing box **is** clean/ and all old packing **has been removed**./
- **Wrap** a turn of packing/ around the shaft/ and **make** it slightly larger/ than the shaft circumference./
- **Cut** the lengths of packing/ with a sharp knife/ to give a clean butt joint./
- **Insert** each turn of packing/ **into** the stuffing box,/ offsetting the butt joints/ by 120°./
- **Put in** sufficient turns/ so that the gland ring nuts **will fit** fully/ **on** the studs./

【専門英語単語リスト】書いておぼえよう

発音したり，スペル練習したり，意味を調べたりしたらチェック欄の○をぬりなさい。
辞書を引いて単語の意味が多くある場合，英語の文脈から判断し，適切な意味を選びなさい。

英文	単語	品詞	意味を調べて書きなさい	発音しながらスペル練習しなさい	チェック
1)	gland	名詞			○○○
2)	seal	名詞			○○○
2)	with great care	熟語			○○○
3)	removal / remove	名 / 動			○○○
4)	fluid	名詞			○○○
5)	avoid Ving	他動詞			○○○
5)	wipe	他動詞			○○○
5)	scratch	名詞			○○○
6)	conventional	形容詞			○○○
6)	replace	他動詞			○○○
8)	circumference	名詞			○○○
9)	length / long	名 / 形			○○○
9)	butt joint	名詞			○○○
10)	offset	他動詞			○○○
11)	nut / stud	名詞			○○○

Unit 32　発電機

Step Check Box □ □ □

Step 1　下記の英文を読み，次の3つの活動をしなさい。終了後 Step Check Box □をぬりなさい。

(1) 動詞（述語動詞）を○で囲む。動詞の意味を考えて，動詞のあとに続く内容を推測する。
(2) 主語（主部）に下線を引く。
(3) スラッシュを入れる。スラッシュを入れる場所は主に下記の3つ。
- コンマ (,) やピリオド (.) でスラッシュ (/) を入れる。音読するときは，スラッシュで息つぎする。
- 接続詞の前にスラッシュを入れる。［接続詞＋主語＋動詞］は文としてまとまった意味を表す。
  ここでいう接続詞とは when, if, until, before, after など（and, but, or, so を含まない）。
- 原則として前置詞の前にスラッシュを入れる。（前置詞＋名詞）は，最小の意味のまとまりを表す。

1) Generators change mechanical energy into electrical energy.
2) Motors change electrical energy into mechanical energy.
3) Generators and motors are a lot alike.
4) They are made in the same general way.
5) Further, they both depend on the same electromagnetic principles for their operation.
6) The first principle is called GENERATOR ACTION.
7) It is also called INDUCTION.
8) Voltage can be induced into a wire that is in a magnetic field.
9) This happens when the magnetic flux is cut by the wire. In some cases, the wire moves.
10) In other cases, the field moves.
11) In still others, both are moving, but at different speeds.
12) It takes mechanical energy to cause the motion.
13) The motion causes electricity to be generated.
14) The second principle is called MOTOR ACTION.
15) This is simply the mechanical forces between magnets.
16) When two magnets (or electromagnets) approach each other, one will be either pulled toward or pushed away from the other.

出典：2級海技士（機関）平成24年4月定期試験問題

Step 2　網掛けは主語，**ゴシック体**は動詞（述語動詞），＊語注・文法的留意事項，①②③は訳順番号

(1) 単語や熟語の説明を読み，わからない単語の意味を調べなさい。
(2) ①②③の訳順番号を参考に，全文和訳を書きなさい。
　　全文和訳が終わったら，Step Check Box □をぬりなさい。

1) ①Generators ④**change** ②mechanical energy/ ③into electrical energy./

- ＊generator：〔名詞〕発電機
- ＊mechanical：〔形容詞〕機械的な
- ＊mechanical energy：機械エネルギー，力学エネルギー
- ＊electrical energy：電気エネルギー
- ＊electricity：〔名詞〕電気，電力，電流
- ＊change A into B：〔他動詞〕AをBに変える
- ＊mechanics：〔名詞〕力学，機械学
- ＊electrical：〔形容詞〕電気による，電気の
- ＊主語が generator なので，「発電機の仕組み」が問われていることが推察できる。主語＋動詞の意味を理解することが大切である。

全文和訳

2) ①Motors ④**change** ②electrical energy/ ③into mechanical energy./

- ＊motor：〔名詞〕電動機，発動機
- ＊change A into B：AをBに変える

Unit 32 発電機

全文和訳

3) Generators and motors **are** a lot alike./
    ① ④ ② ③

* a lot：たくさん，多いに（副詞的に動詞や副詞や形容詞にかかる）
* alike：〔形容詞〕よく似ていて，同様で，等しい
* are：動詞がbe動詞である。「AはBである」と訳す。Bは形容詞あるいは名詞である。

全文和訳

4) They **are made**/ in the same general way./
    ① ③ ②

* 主語theyは3)のGenerators and motorsを指す。
* in the same general way：同じような一般的な方法で

全文和訳

5) Further,/ they both **depend on** the same electromagnetic principles/ for their operation./
    ① ④ ③ ②

* further：〔副詞〕さらに，それ以上に = moreover   * they：3)のgenerators and motorsを指す。
* depend on：〔熟語〕〜による，〜に依存する。②③④の大意は「作動するには〜の原理による」となる。
* electromagnetic：〔形容詞〕電磁石の，電磁気の（略語：E M, e.m.）   * principle：〔名詞〕原理
* ③は，6)のthe first principleにつながる。
* operation：〔名詞〕（機械などの）運転，操作 → operate：〔他動詞〕
* for their operation：作動するために

全文和訳

6) The first principle **is called** GENERATOR ACTION./
    ① ③ ②

* the first principle：第1の原理   * generator action：〔名詞〕発電機の働き

全文和訳

7) It **is also called** INDUCTION./
    ① ③ ②

* induction：〔名詞〕誘導   * induce：〔他動詞〕（電気，磁気などを）誘導する

全文和訳

8) Voltage **can be induced**/ into a wire/ [that **is** in a magnetic field./]
    ① ④ ③ ②

* voltage：〔名詞〕電圧，電位差，ボルト数   * can be induced into 〜：〜へと誘導される
* wire：〔名詞〕電線，針金   * that：〔関係代名詞〕直前の名詞wireにかかる。
* be in a magnetic field：磁場のなかにある   * magnetic：〔形容詞〕磁気の
* 大意：誘導電気は磁界にある電線に電圧を誘導することができる（能動態で考えると意味がわかりやすい）
   = Induction can induce voltage into a wire [that is in a magnetic field.]

全文和訳　電圧は，

135

9) This **happens**/ [when the magnetic flux **is cut**/ by the wire./]
　　④　　　　　　　①　　　　　　　　　③　　　②

* ①②③の大意：磁束が電線で切られるときに
* magnetic flux：磁束
* when 以下（接続詞節という）は動詞 happen にかかる
* flux：〔名詞〕流れ，流動

全文和訳

10) In some cases,/ the wire **moves**./ In other cases,/ the field **moves**./
　　　③　　　　　　①　　②　　　　⑥　　　　　　④　　⑤

* In some cases ～．In other cases ～．：～の場合もあれば，～の場合もある
* field：〔名詞〕（電気的・磁気的に力を及ぼす）場，（粒子の）場，界磁

全文和訳

11) In still others,/ both **are moving**,/ but at different speeds./
　　　①　　　　　③　　　④　　　　　　②

* in still others：ほかの場合には
* but at different speeds：異なる速度であるが
* both：ここでは，10) の wire と field を指す。

全文和訳

12) It **takes** mechanical energy/ to cause the motion./
　　①　④　　　③　　　　　　②

* It takes：It (= the first principle) は～を必要とする（用いる）
* mechanical energy：機械エネルギー，力学エネルギー
* to cause the motion：運動を引き起こすために
* cause：〔動詞〕～を引き起こす
* motion：運動，運転

全文和訳

13) The motion **causes** electricity/ to be generated./
　　　①　　　　④　　　②　　　　　③

* cause ＋目的語＋ to V：〔動詞〕～に～するようにさせる
* to be generated：〔不定詞の受動態〕能動態のように訳す。「運動が電気を発生する」となる。

全文和訳

14) The second principle **is called** MOTOR ACTION./
　　　　①　　　　　③　　　②

* the second principle：第 2 の原理
* motor action：電動機作用

全文和訳

15) This **is** simply the mechanical forces/ between magnets./
　　①　⑤　②　　　　④　　　　　　③

* simply：〔副詞〕単に
* between magnets：磁石の間の （これは the mechanical forces にかかる）
* mechanical force：機械力

全文和訳

16) [When two magnets (or electromagnets) **approach** each other,]/ one **will be** either **pulled** toward (the other)/ or **pushed** away/ from the other./

* ⑤⑥⑦⑧⑨⑩の大意：一方が他方を吸引するか，あるいは他を反発する
* approach：〔他動詞〕～に接近する　　　　　　　　　　* each other：〔熟語〕お互いに
* one ～ the other ～：（慣用的な表現）（2つあるものの）1つは～で，もう片方は～である
* either A or B：（慣用的な表現）Aであるか，あるいはBであるか
* be pulled toward ～：～に向かって吸引される　　　* toward：〔前置詞〕～の方向に（向かって）
* be pushed away from ～：～から反発される

全文和訳

Step 3　網掛け は主語，ゴシック体 は動詞（述語動詞）

(1) 音声を聞きながら，音読する。音読が終わったら，Step Check Box □をぬりなさい。
(2) 左から順に，スラッシュ毎の意味を考える。

Generators **change** mechanical energy/ into electrical energy./ Motors **change** electrical energy/ into mechanical energy./
Generators and motors **are** a lot alike./ They **are made**/ in the same general way./ Further,/ they both **depend on** the same electromagnetic principles/ for their operation./
The first principle **is called** GENERATOR ACTION./ It **is also called** INDUCTION./ Voltage **can be induced**/ into a wire/ that **is** in a magnetic field./ This **happens**/ when the magnetic flux **is cut**/ by the wire./
In some cases,/ the wire **moves**./ In other cases,/ the field **moves**./ In still others,/ both **are moving**,/ but at different speeds./ It **takes** mechanical energy/ to cause the motion./ The motion **causes** electricity/ to be generated./
The second principle **is called** MOTOR ACTION./ This **is** simply the mechanical forces/ between magnets./ When two magnets (or electromagnets) **approach** each other,/ one **will be** either **pulled** toward/ or **pushed** away/ from the other./

【専門英語単語リスト】書いておぼえよう

英文	単語	品詞	意味を調べて書きなさい	発音しながらスペル練習しなさい	チェック
1)	generator	名詞			○○○
1)	mechanical	形容詞			○○○
1)	electrical / electricity	形 / 名			○○○
5)	depend on	熟語			○○○
5)	electromagnetic	形容詞			○○○
7)	induction / induce	名 / 動			○○○
8)	magnetic field	名詞			○○○
9)	flux	名詞			○○○
9)	move / motion	動 / 名			○○○
13)	generate / generation	動 / 名			○○○
15)	force	名詞			○○○
16)	approach	他動詞			○○○

理解を深めるキーワード「発電機の原理」

発電機と電動機は概ね構造が一緒である。

電動機では磁界(Field)のなかに導体(Conductor)を置き，これに電流(Electric current)を流すと，フレミングの左手の法則に従い導体に力を生じる。これを電動機作用(Motor action)という。これにより電気から動力が得られる。

他方，発電機ではフレミングの右手の法則に従う。固定磁界のなかに導体を置き，これを外部の動力で磁界を切るように動かすと，導体に電流が流れる。これを発電機作用(Generator action)または電磁誘導(Electromagnetic induction or Induction)という。

また，前述の発電機作用の磁界[*1]と導体[*2]の関係において，磁界を回転し，導体を固定しても両者の相対関係は変わらず，発電機作用が生じる。

前者を回転電機子型(Revolving-armature type)，後者を回転界磁型(Revolving-field type)という。

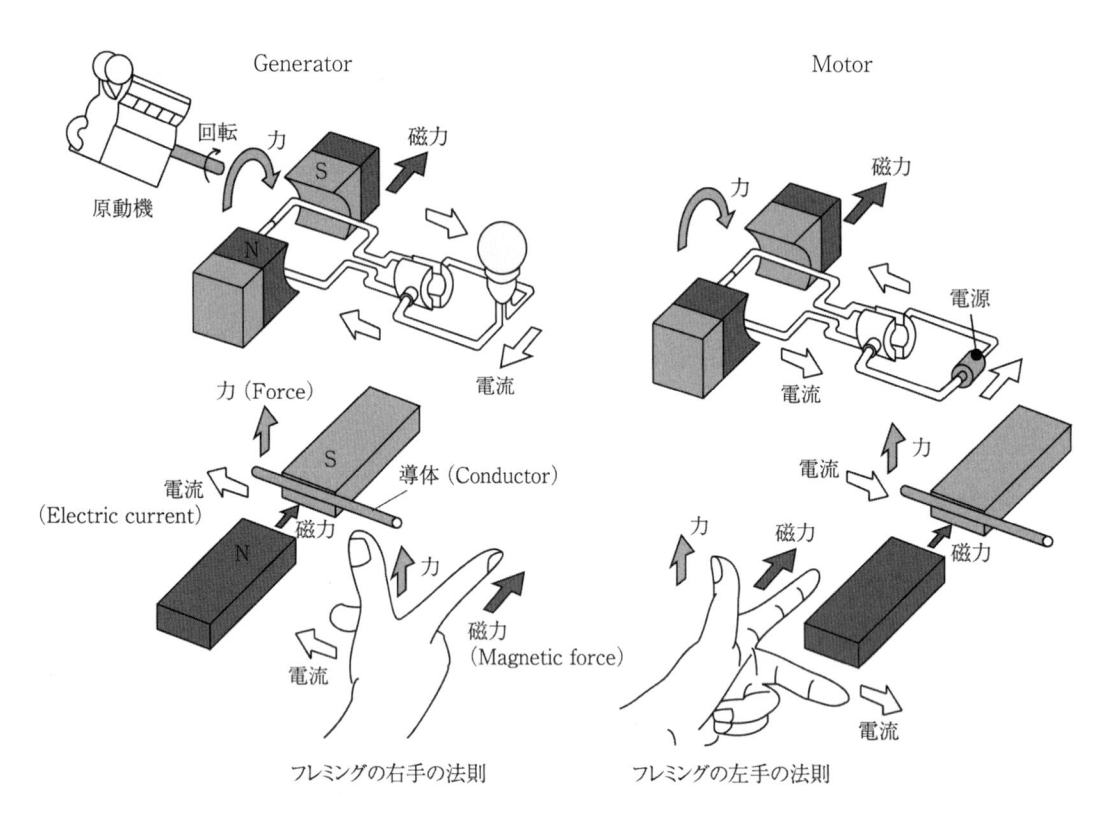

＊1：磁界を発生させる部分を界磁(Field)
＊2：誘導起電力を生じる部分を電機子(Armature)

コラム3

前置詞1

6ページの表を見てください。これはこの本の英文の単語の登場回数のランキングです。ship という単語が9位にランクインしているところなどは，この本らしいところですね。さて，ランキングのなかで 網掛け になっている単語は前置詞という種類の単語です。前置詞は数が少ない上に形が変化しないので，覚えておくと英文を読むヒントになります。

前置詞に関してはまず次の3つのことを覚えておきましょう。

1.「前置詞」の後ろには「名詞（ヒト・モノ）を中心とする意味のまとまり」が来る。
2. 英文の意味がよくわからないときは，1. の（前置詞＋名詞）の部分は無視して英文の意味を考えて，（前置詞＋名詞）の部分の意味は後で付け足す。
3.（前置詞＋名詞）は直前の意味のまとまりか，動詞につながる。

それでは次の英文を読んでみましょう。

The officer in charge of the navigaional watch shall take frequent and accurate compass bearings of approaching ships as a means of early detection of risk of collision.

上の英文の（前置詞＋名詞）を（　）で囲むと

The officer (in charge) (of the navigational watch) shall take frequent and accurate compass bearings (of approaching ships) (as a means) (of early detection) (of risk) (of collision).

次に，（前置詞＋名詞）はとりあえず無視して英文の意味を考えてみます。

The officer shall take frequent and accurate compass bearings.
オフィサーは頻繁で正確なコンパス方位を測らなければならない。

最後に，最初に無視した（前置詞＋名詞）の部分の意味を付け足していきます。

（航海当直の）（担当の）オフィサーは（衝突の）（危険の）（初期の探知の）（手段として）（近づいてくる船の）頻繁で正確なコンパス方位を測らなければならない。

ちょっと英文を読むのが楽になったと思いませんか？

Unit 33　単相と三相

Step Check Box □ □ □

Step 1　下記の英文を読み，次の3つの活動をしなさい。終了後 Step Check Box □をぬりなさい。

(1) 動詞（述語動詞）を○で囲む。動詞の意味を考えて，動詞のあとに続く内容を推測する。
(2) 主語（主部）に下線を引く。
(3) スラッシュを入れる。スラッシュを入れる場所は主に下記の3つ。
   - コンマ (,) やピリオド (.) でスラッシュ (/) を入れる。音読するときは，スラッシュで息つぎする。
   - 接続詞の前にスラッシュを入れる。［接続詞＋主語＋動詞］は文としてまとまった意味を表す。
     ここでいう接続詞とは when, if, until, before, after など（and, but, or, so を含まない）。
   - 原則として前置詞の前にスラッシュを入れる。（前置詞＋名詞）は，最小の意味のまとまりを表す。

1) Household power is single-phase.
2) In fact, most of the lights and other electrical equipment we use every day operate on single-phase ac.
3) Heavy users of power, however, operate with three-phase current.
4) You will normally find three-phase in commercial buildings, manufacturing plants and retail stores.
5) Huge alternators at power generating plants produce three-phase power.
6) It has several advantages:
7) It is cheaper to transmit than single phase.
8) Three-phase motors are simpler, less expensive and more powerful.
9) ALTERNATORS
10) There are two general types of alternators:
11) The REVOLVING-ARMATURE type. It has a stationary magnetic field and rotating alternating current windings.
12) The REVOLVING-FIELD type. Armature is stationary and field windings turn.

出典：2級海技士（機関）平成24年7月定期試験問題

Step 2　網掛けは主語，ゴシック体は動詞（述語動詞），＊語注・文法的留意事項，①②③は訳順番号

(1) 単語や熟語の説明を読み，わからない単語の意味は調べなさい。
(2) ①②③の訳順番号を参考に，全文和訳を書きなさい。
　　全文和訳が終わったら，Step Check Box □をぬりなさい。

1) ①Household power ③**is** ②single-phase./

　＊household power：家庭用電力
　＊single-phase：（システム，回路，装置などが）単相の，単相交流の

全文和訳

2) ①In fact,/ ③most of the lights and other electrical equipment/ ④[(that) we **use** every day]/ ②**operate on** single-phase ac./ ⑥ ⑤

　＊in fact：〔熟語〕実際は，事実上は　　　　　　＊most of ～：たいていの
　＊light：〔名詞〕明り　　　　　　　　　　　　＊other electrical equipment：他の電化製品（機器）
　＊[(that) we use every day]：〔関係代名詞 that（省略されている）～］の文は，直前の名詞 most of the lights and other electrical equipment にかかる。
　＊operate：〔自動詞〕（機械が）作動する，機能する　　＊operate on：～に作用する，はたらく
　＊single-phase：〔形容詞〕単相の，単相交流の　　　　＊ac：（alternating current の略）交流

全文和訳

3) Heavy users of power,/ however,/ **operate**/ with three-phase current./

* heavy users of power：電力を大量に使用するもの
* with three-phase current：三相交流を使って
* 三相交流とは，互いに120°ずつの角をなし，まったく同じ3つのコイルを一様な磁場のなかで回転させると，位相が互いに120°ずつずれた3組の正弦波の交流が得られること。単相交流より電力輸送効率・利用効率がよく，発電所の発電，送電，誘導電動機その他の電源として用いられる。

全文和訳

4) You **will** normally **find** three-phase/ in commercial buildings, manufacturing plants and retail stores./

* normally：〔副詞〕ふつう
* commercial building：〔名詞〕商業用建物
* manufacturing plant：〔名詞〕製造工場
* retail store：〔名詞〕小売店

全文和訳

5) Huge alternators/ at power generating plants **produce** three-phase power./

* huge：〔形容詞〕巨大な
* alternator：〔名詞〕交流発電機
* power generating plant：発電所
* produce：〔他動詞〕～を生ずる，産する，製造(生産)する

全文和訳

6) It **has** several advantages:/

* it：〔代名詞〕前出の単数形 three-phase power を指す。
* several：〔形容詞〕いくつかの(3～6の数を指す)
* advantage：〔名詞〕利点，長所 ↔ disadvantage

全文和訳

7) It **is** cheaper to transmit/ than single phase./

* it：〔代名詞〕前出の単数形 three-phase power を指す。
* ～ is cheaper than ～：比較級 cheaper なので，ここでは三相(three phase)と単相(single phase)を比べている。
* transmit：〔他動詞〕(電気，熱などを)伝える，送る
* to transmit は不定詞で，「～するのに」と訳す。

全文和訳

8) Three-phase motors **are** simpler, less expensive and more powerful./

* simpler, less expensive and more powerful：比較級なので，単相(single phase)と比べている。
  形容詞の変化：simple-simpler-simplest, little-less-least, much-more-most
* less expensive：あまり単価がかからない
* motor：〔名詞〕原動機，発動機

全文和訳

9) ALTERNATORS　和訳：交流発電機
* 交流発電機は，水力，火力，原子力などのエネルギーを交流電力(正弦波交流)に変換する発電装置。直流発電機のような整流の問題がなく，その出力を変圧器により昇圧して長距離送電が可能なので，大容量の発電機はほとんど交流発電機である。

10) There **are** two general types of alternators:/
　　　　　②　　①

全文和訳

11) The REVOLVING-ARMATURE type./ It **has** a stationary magnetic field/
　　　　　　①　　　　　　　　　　　　　④　　　　　②
and rotating alternating current windings./
　　　　　　③

* revolving：〔形容詞〕回転する，回転装置の
* stationary：〔形容詞〕静止した，動かない
* rotating：〔形容詞〕回転している
* winding：〔名詞〕巻き線
* armature：〔名詞〕電機子，接極子
* magnetic field：磁場，磁界
* alternating current：交流

全文和訳

12) The REVOLVING-FIELD type./ Armature **is** stationary/ and field windings **turn**./
　　　　①　　　　　　②　　③　　　　④　　　　　　　　　　⑤

* field：〔名詞〕界磁，界磁石
* field winding：界磁巻線

全文和訳

Step 3   網掛け は主語，**ゴシック体**は動詞（述語動詞）

(1) 音声を聞きながら，音読する。音読が終わったら，Step Check Box □をぬりなさい。
(2) 左から順に，スラッシュ毎の意味を考える。

　Household power **is** single-phase./ In fact,/ most of the lights and other electrical equipment/ we **use** every day/ **operate on** single-phase ac./ Heavy users of power,/ however,/ **operate**/ with three-phase current./ You **will** normally **find** three-phase/ in commercial buildings, manufacturing plants and retail stores./
　Huge alternators/ at power generating plants **produce** three-phase power./ It **has** several advantages:/
1. It **is** cheaper to transmit/ than single phase./
2. Three-phase motors **are** simpler, less expensive and more powerful./
ALTERNATORS
　There **are** two general types of alternators:/
1. The REVOLVING-ARMATURE type./ It **has** a stationary magnetic field/ and rotating alternating current windings./
2. The REVOLVING-FIELD type./ Armature **is** stationary/ and field windings **turn**./

【合格のポイント】
- 海技士試験は，辞書を持ち込める。日ごろの英語の勉強でも，面倒がらずに，辞書を引く練習をする。
- 英語では，第1文がトピックを提示する。今回の第1文は Household power is single-phase. であり，この英文から，電流についての知識を問う問題であるとわかる。主語 Household power で始まる英文だが，第3文では however（しかしながら）という副詞が挿入されている。これは，主題を否定している単語である。第1文の主語 Household power から，第3文では Heavy users of power が主語に変わり，トピックが電流の Heavy users に移行していく。電気に関する専門知識があれば一層わかりやすくなる。

【専門英語単語リスト】書いておぼえよう

英文	単語	品詞	意味を調べて書きなさい	発音しながらスペル練習しなさい	チェック
1)	household power	名詞			○○○
1)	single-phase	形容詞			○○○
2)	equipment / equip	名／動			○○○
2)	electrical	形容詞			○○○
3)	three-phase current	名詞			○○○
4)	manufacturing plant	名詞			○○○
5)	alternator / alternate	名／動			○○○
5)	produce / product	動／名			○○○
7)	transmit / transmission	動／名			○○○
11)	revolving	形容詞			○○○
11)	armature	名詞			○○○
11)	rotate (→ rotating)	他動詞			○○○
11)	alternating current	名詞			○○○
12)	stationary / state	形／名			○○○
12)	field winding	名詞			○○○

理解を深めるキーワード「単相と三相の違い」

電力の送り方には、単相(Single phase)と三相(Three phase)がある。通常、家庭用電源(Household power)は配線が少なく、電圧が低くて安全な単相が使われ、商用や工場用などでは効率が良くて、単相に比べて$1/\sqrt{3}$の電流で同様の電力が得られ、かつ、送電ロスが少なく設備も小さくできる三相が用いられる。

また、家庭用電源といえども、各家庭に引き込む手前の電柱までは三相(6600V)である。これを引き込み線の手前でトランス(Transformer)により100Vまたは200Vに変圧している。

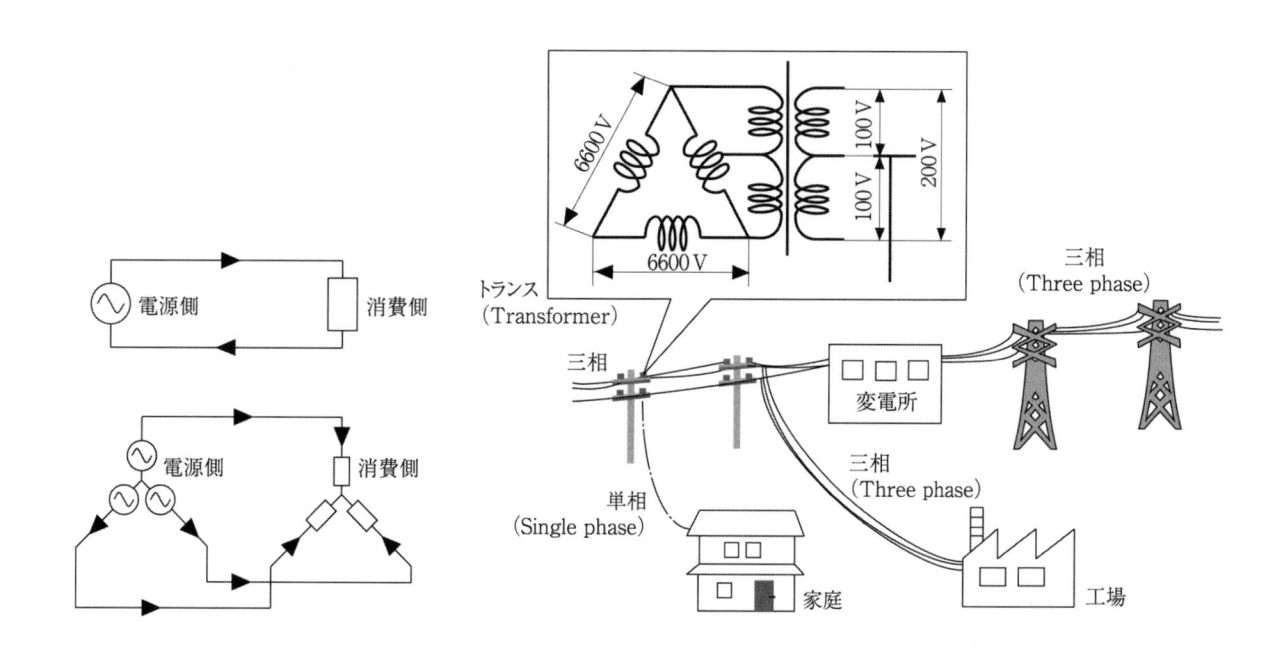

Unit 34　サーキットブレーカー

Step Check Box □ □ □

Step 1　下記の英文を読み，次の3つの活動をしなさい。終了後 Step Check Box □をぬりなさい。

(1) 動詞（述語動詞）を○で囲む。動詞の意味を考えて，動詞のあとに続く内容を推測する。
(2) 主語（主部）に下線を引く。
(3) スラッシュを入れる。スラッシュを入れる場所は主に下記の3つ。
- コンマ (,) やピリオド (.) でスラッシュ (/) を入れる。音読するときは，スラッシュで息つぎする。
- 接続詞の前にスラッシュを入れる。[接続詞＋主語＋動詞] は文としてまとまった意味を表す。
  ここでいう接続詞とは when, if, until, before, after など（and, but, or, so を含まない）。
- 原則として前置詞の前にスラッシュを入れる。(前置詞＋名詞) は，最小の意味のまとまりを表す。

1) Question: What is a circuit breaker? Describe its operation.
2) Answer:
3) A circuit breaker is a mechanical safety device which limits the current flow.
4) The essential part is an electromagnet.
5) The coil becomes energized by the current flowing through the line.
6) When the current in the line reaches a predetermined unsafe value, the coil will develop a magnetic field strong enough to draw up the iron core.
7) When this happens, a latch is tripped and the spring pulls out the breaker arm.
8) Thus the circuit is opened so that no more current can flow.
9) The distance that the iron core can move is adjusted by an adjusting screw.

出典：2級海技士（機関）平成23年2月定期試験問題

Step 2　網掛けは主語，ゴシック体は動詞（述語動詞），＊語注・文法的留意事項，①②③は訳順番号

(1) 単語や熟語の説明を読みながら，部分訳を書きなさい。わからない単語の意味は調べなさい。
(2) 部分訳が終わったら，①②③の訳順番号を参考に，全文和訳を書きなさい。
全文和訳が終わったら，Step Check Box □をぬりなさい。

1) Question: What **is** a circuit breaker?/ **Describe** its operation./

   ＊ circuit breaker：〔名詞〕サーキットブレーカー，回路遮断器
   ＊ describe：〔他動詞〕〜を記述する（説明する）→命令文なので「〜しなさい」と訳す。
   ＊ operate：〔他動詞〕〜を作動する　　　＊ operation：〔名詞〕操作　　　＊ its：それの　　＊ it-its-it

   全文和訳　設問：_____

   ＊設問が「回路遮断器の操作を説明せよ」という意味だから，構造や操作を説明できる力があれば，英文理解が早くなる。

2) Answer　和訳：解答

3) A circuit breaker **is** a mechanical safety device/ [which **limits** the current flow.]/
   　　①　　　　　　④　　　　　　③　　　　　　　　　　　　　　　②

   ＊ mechanical：〔形容詞〕自動的な，機械的な　　＊ mechanic〔名詞〕機械工，熟練工　　＊ mechanics〔名詞〕機械学，力学
   ＊ machine：〔名詞〕機械
   ＊ safety：〔名詞〕安全　　　　　　　　　　　　　＊ safe：〔形容詞〕安全な ↔ unsafe：〔形容詞〕危険な，安全でない
   ＊ device：〔名詞〕装置　　　　　　　　　　　　　＊ limit：〔他動詞〕〜を制限する，限定する
   ＊ current：〔名詞〕電流，流れ
   ＊ flow：〔名詞〕流れ，流れの状態，〔自動詞〕〜が流れる
   ＊関係代名詞 which 以下が device を修飾している。
   ＊関連する単語は，そのなかのベースとなる語を1つ覚えておくと，関連する語の意味を推測しやすくなる。

144

全文和訳

4) The essential part **is** an electromagnet./
  ①   ③   ②

* essential：〔形容詞〕（絶対に）必要な
* electromagnet：〔名詞〕電磁石
* electro-：電気に関する
* magnet：〔名詞〕磁石
* magnetic：〔形容詞〕磁気の

全文和訳

5) The coil **becomes** energized/ by the current/ flowing through the line./
  ①   ②   ④   ③   ②

* coil：〔名詞〕コイル
* energize：〔他動詞〕励磁する，電圧を加える
* becomes energized：was energized のように「電圧を加えられる」と受け身の意味になる。
* current：〔名詞〕電流
* flowing：〔現在分詞〕ここでは，名詞（the current）を修飾する。「〜している」と進行形のように訳す。
* through：〔前置詞〕〜を通って
* line：〔名詞〕電線（回路）

全文和訳

6) When the current in the line **reaches** a predetermined unsafe value,/
     ④   ①   ③   ②

* when：〔接続詞〕〜のとき
  接続詞 when の後に主語 S ＋動詞 V が続く。
* reach：〔他動詞〕達する
* unsafe value：危険な値
* predetermined：〔過去分詞〕あらかじめ決められている

①②③④部分訳 ＿＿＿＿＿＿＿＿＿＿＿＿＿＿＿＿＿＿＿＿＿＿＿＿＿＿

the coil **will develop** a magnetic field/ strong enough to draw up the iron core./
  ⑤   ⑨   ⑧   ⑦   ⑥

* will develop：〔他動詞〕〜をつくりあげる
* magnetic field：〔名詞〕磁界
* draw up：〔他動詞〕引き上げる
* iron core：鉄心
* 形容詞 strong ＋ enough to V（不定詞）：不定詞 to draw up は enough と一緒に使われて，strong にかかる。「〜するのに十分に強い」と訳す。strong ＋ enough to draw up は名詞⑧にかかる（を修飾する）。

⑥⑦⑧部分訳 ＿＿＿＿＿＿＿＿＿＿＿＿＿＿＿＿＿＿＿＿

全文和訳

7) [When this **happens**],/ a latch **is tripped**/ and the spring **pulls out** the breaker arm./
     ①   ②   ③   ⑤   ④

* latch：〔名詞〕掛け金（ばね式の錠前）
* trip：〔自動詞〕（電気回路の一部が安全のために自動的に）切れる，（止め金などが）はずれる → is tripped：〔受動態〕
* spring：〔名詞〕ばね
* pull out：〔熟語〕引き抜く
* breaker arm：〔名詞〕ブレーカーアーム

全文和訳

8) Thus the circuit is opened/ [so that no more current can flow.]/
　　①　　②　　④　　　　　　　③

* thus：〔副詞〕このように
* no more current：名詞 (current) に否定語 (no more) がつくと，日本語に訳すときは，動詞を打ち消しにする。
* no more：「それ以上〜ではない」という意味なので，ここでは「それ以上流れ出ることがないように」と訳す。
* so that S + V：so that は接続詞なので，後に主語と述語が続き，「〜するように」という意味を持つ。

全文和訳

9) The distance [that the iron core can move]/ is adjusted/ by an adjusting screw./
　　②　　　　　　①　　　　　　　　　④　　　　　　③

* distance：〔名詞〕距離
* the iron core：〔名詞〕鉄心
* that：〔関係代名詞〕that 以下の英文 the iron core can move は，直前の名詞 distance を修飾する。
* move：〔他動詞〕〜を動かす
* movement：〔名詞〕動き，運動
* adjust：〔他動詞〕〜を調整する
* be adjusted：〔受動態〕調整される
* adjusting screw：〔名詞〕調節ねじ

全文和訳

Step 3   網掛けは主語，ゴシック体は動詞（述語動詞）

(1) 音声を聞きながら，音読する。音読が終わったら，Step Check Box □をぬりなさい。
(2) 左から順に，スラッシュ毎の意味を考える。

Question: What **is** a circuit breaker?/ **Describe** its operation./
Answer: A circuit breaker **is** a mechanical safety device/ which **limits** the current flow./ The essential part **is** an electromagnet./ The coil **becomes** energized/ by the current/ flowing through the line./ When the current in the line **reaches** a predetermined unsafe value,/ the coil **will develop** a magnetic field/ strong enough to draw up the iron core./ When this **happens**,/ a latch **is tripped**/ and the spring **pulls out** the breaker arm./

Thus/ the circuit **is opened**/ so that no more current **can flow**./ The distance that the iron core **can move**/ **is adjusted**/ by an adjusting screw./

【文法のポイント】
後置修飾のパタンを理解することがポイントである。ここではポイントを5つに分ける。海技士試験に出てきた英文を例に説明する。スラッシュ毎に訳せば，英文の内容が理解できるので，和訳をする場合は，下記を参考に，日本語らしい語順に直すスキルが必要である。

1. ［接続詞＋主語 S ＋動詞 V…］は，動詞 V にかかる。
2. （前置詞＋名詞（もの・コト））は，直前の名詞や動詞にかかる。
3. ［関係代名詞＋主語 S ＋動詞 V…］は，直前の名詞にかかる。
4. 名詞＋過去分詞 (Ved) の語順の場合，過去分詞 (Ved) は，直前の名詞にかかる。
5. 名詞＋現在分詞 (Ving) の語順の場合，現在分詞 (Ving) は，直前の名詞にかかる。

具体例

◆（前置詞＋名詞（もの・コト））が，直前の名詞や動詞にかかる

例1　The coil **becomes** energized/ by the current/ flowing through the line./

　　＊by the current は動詞 becomes にかかる。前置詞 by は「～の近くにいる」「距離感が近い」という意味から「～のそばで」「（そばにある手段や方法）を用いて」「（近くの行為者）によって」と訳す。

◆［関係代名詞＋主語S＋動詞V…］は，直前の名詞にかかる

関係代名詞 which, that は働きが2つある。1つ目は，接続詞として働く。関係代名詞以下の英文のなかには，主語Sと動詞Vが続く。2つ目は，関係代名詞以下の英文が，関係代名詞の直前の名詞にかかる（を修飾する）。

例2　A circuit breaker **is**/ a mechanical safety device/ [which **limits** the current flow.]

　　＊which limits the current flow は直前の名詞 device にかかる。訳は「電流の流れを制限する自動安全装置」となる。

例3　The distance [that the iron core **can move**]/ **is adjusted**/ by an adjusting screw./

　　＊that the iron core can move は，直前の名詞 distance にかかるので，「鉄心が移動する距離」と訳す。

◆（名詞＋現在分詞 Ving）の語順ならば，現在分詞は「～している」と進行形のように訳し，直前の名詞にかかる

例4　The coil **becomes** energized/ by the current/ (flowing through the line.)

　　＊flowing through the line は直前の名詞 the current にかかる。訳は「電流回路を通って流れる電流」となる。

【専門英語単語リスト】書いておぼえよう

発音したり，スペル練習したり，意味を調べたりしたらチェック欄の○をぬりなさい。
辞書を引いて単語の意味が多くある場合，英語の文脈から判断し，適切な意味を選びなさい。

英文	単語	品詞	意味を調べて書きなさい	発音しながらスペル練習しなさい	チェック
1)	circuit breaker	名詞			○○○
3)	mechanical / mechanic	形／名			○○○
3)	safety / safe ↔ unsafe	名／形			○○○
3)	current	名詞			○○○
3)	flow	名／動			○○○
4)	electromagnet	名詞			○○○
5)	energize	他動詞			○○○
6)	reach	他動詞			○○○
6)	develop	他動詞			○○○
6)	magnetic / magnet	形／名			○○○
6)	draw up	他動詞			○○○
7)	trip	他動詞			○○○
7)	pull out	他動詞			○○○
9)	distance	名詞			○○○
9)	adjust	他動詞			○○○

Unit 35　ダイオード

Step Check Box □ □ □

Step 1　下記の英文を読み，次の3つの活動をしなさい。終了後 Step Check Box □をぬりなさい。

(1) 動詞（述語動詞）を○で囲む。動詞の意味を考えて，動詞のあとに続く内容を推測する。
(2) 主語（主部）に下線を引く。
(3) スラッシュを入れる。スラッシュを入れる場所は主に下記の3つ。
- コンマ (,) やピリオド (.) でスラッシュ (/) を入れる。音読するときは，スラッシュで息つぎする。
- 接続詞の前にスラッシュを入れる。［接続詞＋主語＋動詞］は文としてまとまった意味を表す。
  ここでいう接続詞とは when, if, until, before, after など (and, but, or, so を含まない)。
- 原則として前置詞の前にスラッシュを入れる。(前置詞＋名詞) は，最小の意味のまとまりを表す。

1) A RECTIFIER DIODE is similar to a single-pole switch.
2) It is a two-terminal device consisting of an anode (made of P-type material) joined to a cathode (made of N-type material).
3) Unlike a simple switch, however, a rectifier diode will conduct current in only one direction.
4) The diode only conducts when its anode is more positive than its cathode.
5) This condition is called FORWARD BIAS, and the resulting current is called *forward current*.
6) When the diode's anode is more negative than its cathode (REVERSE BIAS), the diode will not conduct current.
7) Forward bias voltage for a silicon diode must be greater than 0.6 volts for the diode to switch ON.

出典：2級海技士（機関）平成24年2月定期試験問題

Step 2　網掛け は主語，**ゴシック体**は動詞（述語動詞），＊語注・文法的留意事項，①②③は訳順番号

(1) 単語や熟語の説明を読み，わからない単語の意味は調べなさい。
(2) ①②③の訳順番号を参考に，全文和訳を書きなさい。
全文和訳が終わったら，Step Check Box □をぬりなさい。

1) ① ③ ②
   A RECTIFIER DIODE **is** similar to a single-pole switch./
   * rectifier diode：〔名詞〕整流ダイオード　　　　　＊ diode：〔名詞〕二極真空管，二極体，ダイオード
   * rectifier：〔名詞〕整流器（交流を直流に変えるため，一方向に電流がよく流れるようにする装置）
     交流から直流を得ることを整流といい，整流器はそのための電気的な素子。一方向（順方向）へは電流をよく通し，その反対方法（逆方向）へは電流を通さない。以前は二極電子管などの整流管が用いられたが，近年はシリコンなどの半導体を用いた各種のダイオードが使われる。
   * be similar to 名詞：〔熟語〕～と似ている　　　＊ single-pole switch：〔名詞〕単極スイッチ

全文和訳

2) ① ⑦ ⑥ ⑤ ④ ③ ②
   It **is** a two-terminal device／consisting of an anode／made of P-type material／joined to a cathode／made of N-type material.／
   * terminal：〔名詞〕（電池の）端子，電極　　　　　　＊ device：〔名詞〕装置
   * a two-terminal device：2つのターミナルデバイス（ここではP型半導体とN型半導体を指す）
   * consist of：〔自動詞〕～から成り立つ → 類語：be made up of, be composed of
   * consisting of：〔現在分詞〕直前の名詞 a two-terminal device にかかる
   * anode：〔名詞〕（電子管，半導体などの）陽極，アノード
   * cathode：〔名詞〕（電子管，半導体などの）陰極，カソード
   * P-type material：〔名詞〕P型半導体　　　　　　　＊ N-type material：〔名詞〕N型半導体
   * made of：〔過去分詞〕～でつくられる　　＊ made of P-type material は anode にかかる。過去分詞は受け身に訳す。
   * joined to：〔過去分詞〕～に連結した　　＊ joined to a cathode made of N-type material は anode にかかる。

Unit 35 ダイオード

全文和訳

3) Unlike a simple switch,/ however,/ a rectifier diode **will conduct** current/ in only one direction./
②　　　　　　　①　　　　③　　　　　　⑤　　　　　④

* unlike：〔前置詞〕～と違って
* however：〔副詞〕しかしながら
* rectifier diode：〔名詞〕整流ダイオード
* conduct：〔他動詞〕（金属などが熱や電気を）伝える，伝導する
* current：〔名詞〕電流
* in only one direction：一方向だけで

全文和訳

4) The diode only **conducts**/ [when its anode **is** more positive/ than its cathode./]
①　　　⑥　　⑤　　②　　④　　③

* conduct：〔自動詞〕（電気などが）伝わる
* positive：〔形容詞〕正（陽）電気の，正電荷を持った
* more positive than ～：positive の比較級。
* when 以下は，動詞 conducts にかかる。

全文和訳

5) This condition **is called** FORWARD BIAS,/ and the resulting current **is called** forward current./
①　　③　　②　　　　　　④　　　⑥　　⑤

* condition：〔名詞〕状態
* forward bias：〔名詞〕順方向バイアス→半導体回路（素子）で，電流の流れる方向にかけるバイアス
* Bias：〔名詞〕バイアス→所定の動作点を得るためにトランジスタ制御電極などに加える電圧（偏り）
* resulting current：〔名詞〕結果として生じる電流　　* forward current：〔名詞〕順方向電流

全文和訳

6) [When the diode's anode **is** more negative/ than its cathode (REVERSE BIAS),/] the diode **will not conduct** current./
⑤　　①　　④　　②　　③　　⑥　　⑧　　⑦

* negative：〔形容詞〕負の，陰電気の（を生じる），負の電気を帯びた，陰極の
* more negative than：～より負の電荷を帯びるとき→ negative-more negative-most negative
* reverse bias：逆方向バイアス
  ここでは（　　）になっているので「アノード陽極がカソード陰極（つまり逆方向バイアス）より負の電気を帯びるとき」と訳す。
* [when 以下] は述語動詞 will not conduct にかかる。　　*⑥⑦⑧の大意：ダイオードは電流を導通しない

全文和訳

7) Forward bias voltage for a silicon diode **must be** greater/ than 0.6 volts/ for the diode to switch ON./
①　　⑤　　④　　②　　③

* forward bias voltage for a silicon diode：シリコンダイオードのための順方向バイアス電圧
* volt：ボルト（電圧の単位，略語はV）　　* great-greater-greatest
* ④⑤の大意：0.6 ボルト以上の電圧でなければいけない
* for the diode to switch ON：ダイオードが ON になるために
  for the diode は，不定詞 to switch ON の意味上の主語の働きをしている。

全文和訳

149

Step 3 網掛けは主語，**ゴシック体**は動詞（述語動詞）

(1) 音声を聞きながら，音読する。音読が終わったら，Step Check Box □をぬりなさい。
(2) 左から順に，スラッシュ毎の意味を考える。

A RECTIFIER DIODE **is** similar to a single-pole switch./ It **is** a two-terminal device/ consisting of an anode/ made of P-type material/ joined to a cathode/ made of N-type material./ Unlike a simple switch,/ however,/ a rectifier diode **will conduct** current/ in only one direction./ The diode only **conducts**/ when its anode **is** more positive/ than its cathode./ This condition **is called** FORWARD BIAS,/ and the resulting current **is called** *forward current*./ When the diode's anode **is** more negative/ than its cathode (REVERSE BIAS),/ the diode **will not conduct** current./ Forward bias voltage for a silicon diode **must be** greater/ than 0.6 volts/ for the diode to switch ON./

【合格のポイント】
文章の最初の主語が，トピックを表している。ここでは最初の主語が A RECTIFIER DIODE なので，後に続く英文は，それに関連した内容が詳細に具体的に記載されている。英語では，トピックがまず示されると，次にそれについての詳細で具体的な情報が続く。このように英語が書かれていることを知っておくと，長文の流れがわかりやすい。下記のように主語を追っていくと，A RECTIFIER DIODE がトピックとして流れていることがわかる。

A RECTIFIER DIODE **is** … It **is** a two-terminal device … a rectifier diode **will conduct** … The diode only **conducts** … This condition **is called** … the diode **will not conduct** current … Forward bias voltage for a silicon diode **must be** …

【専門英語単語リスト】書いておぼえよう

英文	単語	品詞	意味を調べて書きなさい	発音しながらスペル練習しなさい	チェック
1)	rectifier diode	名詞			○○○
1)	(be) similar to 名詞	形容詞			○○○
1)	single-pole switch	名詞			○○○
2)	consist of	熟語			○○○
2)	anode	名詞			○○○
2)	cathode	名詞			○○○
3)	conduct	他動詞			○○○
3)	current	名詞			○○○
4)	positive	形容詞			○○○
5)	forward bias	名詞			○○○
5)	forward current	名詞			○○○
6)	negative	形容詞			○○○
6)	reverse bias	名詞			○○○
7)	voltage / volt	名詞			○○○
7)	switch on ↔ switch off	熟語			○○○

～理解を深めるキーワード「ダイオード（DIODE）」～

ダイオードについては，Unit 50 にその構造や種類について説明されている。授業で習った知識を活かして熟読しよう。

コラム4

前置詞2　ofの訳し方

6ページの表にあったように，ofは英語のなかでいちばんよく使われる前置詞です。

前置詞なので，後ろにはたいてい「名詞（ヒト・モノ）」が続きますが，ofの場合は前に名詞があるかどうかを確認しましょう。

そして「名詞（前）」of「名詞（後）」という形になっているときは，「名詞（後）の名詞（前）」と訳して意味が通じるかを考えます。

```
英語                                    日本語
round 7 percent of NOx emissions  ⇒  窒素酸化物の放出 の 約7パーセント
   モノ（前）      モノ（後）              モノ（後）        モノ（前）
```

ofがいくつか連続するときも同じように「名詞（前）」of「名詞（後）」ごとに意味を考えてかまいませんが，たいていはいちばん最後の名詞から順番に訳していくことになります。

```
英語                                                  日本語
a means of early detection of risk of collision  ⇒  衝突 の 危険 の 早期の探知 の 手段
 名詞1     名詞2          名詞3    名詞4           名詞4   名詞3    名詞2       名詞1
```

151

Unit 36　排ガス規制

Step Check Box □ □ □

Step 1　下記の英文を読み，次の3つの活動をしなさい。終了後 Step Check Box □ をぬりなさい。

(1) 動詞（述語動詞）を○で囲む。動詞の意味を考えて，動詞のあとに続く内容を推測する。
(2) 主語（主部）に下線を引く。
(3) スラッシュを入れる。スラッシュを入れる場所は主に下記の3つ。
- コンマ(,)やピリオド(.)でスラッシュ(/)を入れる。音読するときは，スラッシュで息つぎする。
- 接続詞の前にスラッシュを入れる。［接続詞＋主語＋動詞］は文としてまとまった意味を表す。
  ここでいう接続詞とは when, if, until, before, after など（and, but, or, so を含まない）。
- 原則として前置詞の前にスラッシュを入れる。（前置詞＋名詞）は，最小の意味のまとまりを表す。

1) Marine engine designers in recent years have had to address the challenge of tightening controls on noxious exhaust gas emissions imposed by regional, national and international authorities responding to concern over atmospheric pollution.

2) Exhaust gas emissions from marine diesel engines largely comprise nitrogen, oxygen, carbon dioxide and water vapour, with smaller quantities of carbon monoxide, oxides of sulphur and nitrogen, partially reacted and non-combusted hydrocarbons and particulate material (Figure 3.1).

3) Nitrogen oxides (NOx) are of special concern since they contribute to photochemical smog formation over cities and acid rain.

4) Sulphur oxides (SOx) increase the acidity of the soil and have detrimental effects on human respiration, vegetation and building materials.

5) International shipping reportedly accounts for 4-5 per cent of SOx emissions and around 7 per cent of NOx emissions worldwide.

出典：1級海技士（機関）平成23年10月定期試験問題

Step 2　網掛けは主語，ゴシック体は動詞（述語動詞），＊語注・文法的留意事項，①②③は訳順番号

(1) 単語や熟語の説明を読みながら，部分訳を書きなさい。わからない単語の意味は調べなさい。
(2) 部分訳が終わったら，①②③の訳順番号を参考に，全文和訳を書きなさい。
　　全文和訳が終わったら，Step Check Box □ をぬりなさい。

1) Marine engine designers/ in recent years **have had to address** the challenge/ of tightening controls/on noxious exhaust gas emissions/ imposed by regional, national and international authorities/ responding to concern/ over atmospheric pollution./

   ＊ marine engine designer：船舶用機関設計者
   ＊ have had to address：(～の課題に) 取り組まなければいけなくなってきている　　＊ have had：現在完了形
   ＊ address：〔他動詞〕(課題などに) 取り組む, (問題などを) 扱う
   ＊ challenge：〔名詞〕やりがいのある課題
   ＊ tighten：〔他動詞〕(規則などを) 厳重にする, 強化する　　＊ tight：〔形容詞〕厳しい, ぴったりと合った
   ＊ control：〔名詞〕制限, 管理, 抑制　　　　　　　　　　　＊ on：〔前置詞〕に関する, について
   ＊ noxious：〔形容詞〕有害な, 健康に悪い＝harmful　　　　＊ exhaust gas：〔名詞〕排気ガス
   ＊ emission：〔名詞〕放出, 排気　　　　　　　　　　　　　＊ emit：〔他動詞〕放つ, ～を出す
   ＊ imposed：〔過去分詞〕～を押しつけられた, 課せられた　　＊ imposed 以下が名詞 (emissions) を修飾する。
   ＊ by regional, national, international authorities：地域的, 国内的, 国際的な機関によって
   ＊ responding to：〔現在分詞〕～に反応している。responding to 以下が authorities を修飾する。
   ＊ concern：〔名詞〕不安, 重大事　　　　　　　　　　　　＊ over：〔前置詞〕～をめぐって, について
   ＊ atmospheric：〔形容詞〕大気の　　＊ atmosphere：〔名詞〕大気　　＊ atmospheric pollution：大気汚染
   ＊ タイトルは付いていないが, 最初の英文から, 排気ガス放出の規制が話題（トピック）になっていることがわかる。英文の構成は, 大きなトピックを passage（一節）の始めに述べ, 次に, 具体的なことが順次, 述べられることが多い。

③④⑤⑥⑦部分訳 _____ についての (制限)

全文和訳

2) Exhaust gas emissions/ from marine diesel engines largely **comprise** nitrogen, oxygen, carbon dioxide and water vapour,/ with smaller quantities of carbon monoxide, oxides of sulphur and nitrogen,/ partially reacted and non-combusted hydrocarbons/ and particulate material (Figure 3.1)./

- * marine diesel engine：船舶用ディーゼル機関
- * comprise：〔他動詞〕～を含む ＝ include, consist of   * largely：〔副詞〕主に
- * nitrogen, oxygen, carbon dioxide：〔名詞〕窒素（N），酸素（O），二酸化炭素（CO$_2$）
- * water vapour：〔名詞〕水蒸気
  vapour はイギリス英語のスペル，アメリカ英語のスペルは vapor
  海技士試験ではイギリス英語のスペルが使用されることが多い。
- * with：〔前置詞〕～を伴って，～と一緒に
  (with ③④⑤) は，直前の名詞⑥⑦⑧⑨にかかる。
- * quantity of ～：〔熟語〕～の分量，数量の
- * oxide：〔名詞〕酸化物   * monoxide：〔名詞〕一酸化物   * dioxide：〔名詞〕二酸化物
- * sulphur：硫黄（S）
  sulphur はイギリス英語のスペル，アメリカ英語のスペルは sulfur
- * oxides of sulphur and nitrogen：硫黄酸化物と窒素酸化物   * partially：〔副詞〕部分的には
- * reacted：〔過去分詞〕反応を起こした→ hydrocarbon にかかる
- * non-combusted：〔形容詞〕未燃焼の→ hydrocarbon にかかる   * hydrocarbon：〔名詞〕炭化水素
- * particulate material：微粒子物質   * Figure 3.1：図 3.1（を参照せよ）
- * 大意：①からの排気ガスの放出物は，③④⑤を伴う⑥と⑦と⑧と⑨から主に成っている

全文和訳

3) Nitrogen oxides (NOx) **are** of special concern/ [since they **contribute to** photochemical smog formation over cities/ and acid rain.]/

- * nitrogen oxide：窒素酸化物（NOx）
- * of concern：〔形容詞〕関心のある
  of ＋名詞 ＝ 形容詞→例：of value ＝ valuable, of use ＝ useful
- * since：〔接続詞〕～なのだから（＝ because）
  since の後には主語 S と動詞 V が続く。
- * contribute to：〔自動詞〕への一因となる，寄与する
- * photochemical smog formation：光化学スモッグ形成
- * form：〔他動詞〕～を形成する，～の構成要素になる
- * over：〔前置詞〕～の上方に，～のすみずみまで，…一面に
  photochemical smog formation にかかる。
- * acid rain：酸性雨   *接続詞 and は③②と④をつなぐ。

全文和訳

4) Sulphur oxides (SOx) **increase** the acidity of the soil/ and **have** detrimental effects/ on human respiration, vegetation and building materials./

- * sulphur oxides：硫黄酸化物（SOx）   * acidity：〔名詞〕酸性度   * acid：〔形容詞〕酸性の
- * soil：〔名詞〕土壌   * detrimental：〔形容詞〕有害な
- * effect (on ～)：〔名詞〕（原因に対する）結果，影響
- * respiration：〔名詞〕呼吸作用   * vegetation：〔名詞〕植物，植生   * vegetable：〔名詞〕野菜
- * building material：建築資材

全文和訳

5) International shipping reportedly **accounts for** 4-5 per cent of SOx emissions/ and around 7 per cent of NOx emissions worldwide./

- * international shipping：外航海運業
- * account for ～：〔熟語〕～に責任を持つ，～の説明となる
- * emission：〔名詞〕
- * NOx：窒素酸化物
- * around 7 per cent of NOx emissions worldwide：世界中の窒素酸化物の放出のおよそ7％
- * reportedly：〔副詞〕報道によれば
- * SOx：硫黄酸化物
- * worldwide：〔形容詞〕世界的な

全文和訳

Step 3　網掛けは主語，ゴシック体は動詞（述語動詞）

(1) 音声を聞きながら，音読する。音読が終わったら，Step Check Box □をぬりなさい。
(2) 左から順に，スラッシュ毎の意味を考える。

Marine engine designers/ in recent years **have had to address** the challenge/ of tightening controls/ on noxious exhaust gas emissions/ imposed by regional, national and international authorities/ responding to concern/ over atmospheric pollution./

Exhaust gas emissions/ from marine diesel engines largely **comprise** nitrogen, oxygen, carbon dioxide and water vapour,/ with smaller quantities of carbon monoxide, oxides of sulphur and nitrogen,/ partially reacted and non-combusted hydrocarbons/ and particulate material (Figure 3.1)./ Nitrogen oxides (NOx) **are** of special concern/ since they **contribute to** photochemical smog formation over cities/ and acid rain./ Sulphur oxides (SOx) **increase** the acidity of the soil/ and **have** detrimental effects/ on human respiration, vegetation and building materials./

International shipping reportedly **accounts for** 4-5 per cent of SOx emissions/ and around 7 per cent of NOx emissions worldwide./

【専門英語単語リスト】書いておぼえよう

発音したり，スペル練習したり，意味を調べたりしたらチェック欄の○をぬりなさい。
辞書を引いて単語の意味が多くある場合，英語の文脈から判断し，適切な意味を選びなさい。

英文	単語	品詞	意味を調べて書きなさい	発音しながらスペル練習しなさい	チェック
1)	marine engine designer	名詞			○○○
1)	address	他動詞			○○○
1)	tighten / tight	動/形			○○○
1)	exhaust gas emission	名詞			○○○
1)	impose	他動詞			○○○
1)	authority	名詞			○○○
1)	respond to 名詞	熟語			○○○
2)	comprise	他動詞			○○○
2)	non-combusted	形容詞			○○○
2)	particulate	形容詞			○○○
3)	of special concern	熟語			○○○
3)	contribute to 名詞	熟語			○○○
4)	effect	名詞			○○○
5)	reportedly	副詞			○○○
5)	account for	熟語			○○○

理解を深めるキーワード「排ガス規制」

　国際間の物資の輸送にたずさわる外航船舶においては，環境汚染にかかわる問題は一国で対処できない性質のものであり，国際的なルールづくりが不可欠となっている。

　国際海事機関（International Maritime Organization：IMO）では，条約や規則を定め，各国にこれを遵守させることで対策を講じようとしている。このなかでも船舶からの排気ガスによる大気汚染はIMOの専門委員会である海洋環境保護委員会（Marine Environment Protection Committee：MEPC）で取り扱われている。

　大気汚染防止については，当初，1997年に採択されたMARPOL（The International Convention for the Prevention of Pollution from Ships）条約附属書Ⅳにより，船舶のディーゼルエンジンからの窒素酸化物や燃料油中の硫黄含有率が規制された。しかし，その後の技術革新や各地域での大気汚染が深刻化したため，規制強化の検討がなされた。

　2007年，IMOにおいて，この規制強化の内容が採択され，2010年7月から新しい規制が開始された。主な改正内容は次のとおり。

1. NOx
   ①新規エンジンの規制
   　排出規制海域内（Emission Control Area：ECA）を航行する船舶に搭載するエンジンは，これまでより80％削減した規制値を適用（24 m以下の船舶に搭載されるものなど一部のエンジンを除く）；2016年1月1日～
   ②既存エンジンの規制
   　1990年代に製造され，かつ5000 kW以上かつ90 L/CYL（一気筒の行程容積）以上のエンジンであって，主管庁が，削減方法があると認めるエンジンには適用する。

2. SOx
   　基本的には次のとおり燃料油中の硫黄分の含有率を削減する方向であるが，同等にSOx分を削減することや発生する汚染水を処理することなどを条件に脱硫装置（De-sulphurization Equipment）の使用を認める。
   ①一般海域
   　2020年 or 2025年　　0.5 ％wt※　　※％wt：重量％
   　2018年　　　　　　　見直し実施
   ②排出規制海域（ECA）
   　2010年7月～　　　　1.0 ％wt
   　2015年～　　　　　　0.5 ％wt

Unit 37　主機の据付け

Step 1　下記の英文を読み，次の3つの活動をしなさい。終了後 Step Check Box □をぬりなさい。

(1) 動詞（述語動詞）を○で囲む。動詞の意味を考えて，動詞のあとに続く内容を推測する。
(2) 主語（主部）に下線を引く。
(3) スラッシュを入れる。スラッシュを入れる場所は主に下記の3つ。
- コンマ (,) やピリオド (.) でスラッシュ (/) を入れる。音読するときは，スラッシュで息つぎする。
- 接続詞の前にスラッシュを入れる。［接続詞＋主語＋動詞］は文としてまとまった意味を表す。
  ここでいう接続詞とは when, if, until, before, after など (and, but, or, so を含まない)。
- 原則として前置詞の前にスラッシュを入れる。（前置詞＋名詞）は，最小の意味のまとまりを表す。

1) The various kinds of flexible or resilient mounting all serve to keep the engine in the desired position and to prevent transmission of vibration to the hull of the ship.
2) In passenger ships it is essential to avoid vibration and structure borne noise in the accommodation spaces, and but the use of medium speed engines allows passenger spaces to be placed low down in the ship.
3) It is possible to mount engines on a raft with integral vibration damping, but direct mounting of engines in the ship is simpler and less costly.
4) Flexible/resilient mountings are rubber based and may be of the inclined or vertical type.
5) The rubber pads absorb noise and vibration energy.
6) An effective resilient mounting system will provide an isolation factor of at least 50 or 80 per cent.
7) Although rubber has excellent vibration damping properties, it is prone to damage by mineral oil, so the rubber mounts have to be protected against dripping and splashing by means of covers.
8) Brackets for the rubber mounts are fitted on the foundation plate using steel or resin chocks.
9) Buffers, similar to the flexible mountings, are fitted at the sides and ends of the engine.
10) The flexible mountings must be carefully designed to ensure the attenuation of structure borne noise.

（注）structure borne noise：構造伝達騒音　attenuation：減衰

出典：1級海技士（機関）平成24年2月定期試験問題

Step 2　網掛けは主語，ゴシック体は動詞（述語動詞），＊語注・文法的留意事項，①②③は訳順番号

(1) 単語や熟語の説明を読みながら，部分訳を書きなさい。わからない単語の意味は調べなさい。
(2) 部分訳が終わったら，①②③の訳順番号を参考に，全文和訳を書きなさい。
　　全文和訳が終わったら，Step Check Box □をぬりなさい。

1) ① The various kinds of flexible or resilient mounting ② all **serve to keep** ⑤ the engine/ ③ in the desired position/ ④ and (**serve**) **to prevent** ⑦ transmission of vibration/ ⑥ to the hull of the ship./

- ＊ the various kinds of ～：〔熟語〕さまざまな種類の
- ＊ flexible：〔形容詞〕柔軟な，たわみやすい
- ＊ resilient：〔形容詞〕（圧縮に対して）弾力のある
- ＊ mounting：〔名詞〕据付け，取付け
- ＊ serve to 動詞の原形：～するのに役に立つ
- ＊ keep ～ in the desired position：～を望ましい位置に保持する
- ＊ prevent：〔他動詞〕～を防ぐ，～するのを予防する
- ＊ vibrate：〔他動詞〕～を振動させる
- ＊ transmission of vibration：振動の伝達
- ＊ transmit：〔他動詞〕～を伝える，送る
- ＊ the hull of the ship：船の船体（ハル）
- ＊ vibration：〔名詞〕振動

③④⑤部分訳 _____

全文和訳

2) In passenger ships/ it is essential/ to avoid vibration and structure borne noise/ in the accommodation spaces,/ but the use of medium speed engines allows passenger spaces/ to be placed low down/ in the ship./

- ＊ it is essential：it は仮主語で，to avoid vibration 以下が真の主語なので，it は訳さない。
- ＊ essential：〔形容詞〕不可欠な　　　　　　　　　　＊ avoid：〔他動詞〕～を避ける
- ＊ structure borne noise：構造から生まれる騒音＝構造伝達騒音
- ＊ borne：bear-bore-born or borne（過去分詞 borne が名詞 noise にかかる）
- ＊ accommodation space：居住区域
- ＊ the use of medium speed engine：中速機関の使用は→中速エンジンを使えば（使うことは）
- ＊ allows ～ to be placed low down：～を（階）下の低い所に置いてもよい
- ＊ to be placed：不定詞の受動態　　　　　　　　　　＊ allow：〔他動詞〕～であることを認める，許す
- ＊ passenger spaces：〔名詞〕旅客区域　　　　　　　＊ low：〔副詞〕低い所に　　＊ down：〔副詞〕下方へ

②③④⑤部分訳＿＿＿

全文和訳＿＿

3) [It is possible/ to mount engines/ on a raft with integral vibration damping,/] but [direct mounting of engines in the ship is simpler and less costly.]/

- ＊ it is possible：it は仮主語で，to mount engines 以下が真の主語。
- ＊ mount ～ on …：〔他動詞〕…に～を取り付ける　　＊ raft：〔名詞〕浮き台
- ＊ on a raft with integral vibration damping：振動減衰がついた一体型の浮き台の上に
- ＊ integral：〔形容詞〕（全体の一部分として）一体化した，一体をなす　　＊ vibration damping：振動減衰
- ＊ damping：〔名詞〕（振動などの）制動，減衰，ダンピング，エネルギー分散
- ＊ with：〔前置詞〕～がついている　　　　　　　　　＊ with の基本のイメージは「一緒に」。
- ＊ direct mounting of engines：エンジンを直接据え付けること
- ＊ simple-simpler-simplest → simpler（比較級）：より簡単な
- ＊ costly：〔形容詞〕高価な，費用のかかる　　　　　＊ little-less-least
- ＊ less：〔形容詞〕（little の比較級）より少ない，より小さい　　＊ less costly：あまり費用がかからない

①②③部分訳＿＿

⑤⑥部分訳＿＿

全文和訳＿＿

4) Flexible/resilient mountings are rubber based/ and may be/ of the inclined or vertical type./

- ＊ flexible/resilient mountings are：スラッシュは and の意味。
- ＊ flexible：〔形容詞〕弾力的な，柔軟な，たわみやすい　　＊ resilient：〔形容詞〕弾力性のある，弾性のある
- ＊ mounting：〔名詞〕取付け　　　　　　　　　　　　＊ rubber based：〔形容詞〕ゴムを基にした
- ＊ inclined：〔形容詞〕傾斜の，斜めの　　　　　　　＊ vertical：〔形容詞〕垂直の
- ＊（of + 名詞）→ 形容詞の働きをする：（例）of value = valuable

全文和訳＿＿

5) The rubber pads absorb noise and vibration energy./

- ＊ pad：〔名詞〕（円板ブレーキの）摩擦材，船首防衝材，パッド　　＊ absorb：〔他動詞〕～を吸収する

全文和訳＿＿

6) An effective resilient mounting system **will provide** an isolation factor/ of at least 50 or 80 per cent./
   ①effective ④will provide ②an isolation factor ③of at least 50 or 80 per cent

   * effective：〔形容詞〕効果的な
   * isolation：〔名詞〕遮断，隔離
   * at least：〔熟語〕少なくとも
   * provide：〔他動詞〕～をもたらす，用意する，提供する
   * factor：〔名詞〕率，係数，要素
   * little-less-least

全文和訳

7) [Although ④rubber ①has ③excellent vibration damping properties],/ ②it ⑤is prone to damage/ by ⑥mineral oil,/ ⑦so [⑧the rubber mounts ⑪have to be protected/ ⑩against dripping and splashing/ ⑨by means of covers.]/

   * although S + V：〔接続詞〕～けれども = though
   * property：〔名詞〕（ものに固有の）特性，性質
   * be prone to 名詞：（～の好ましくない）傾向がある，～しがちである
   * it is prone to damage by mineral oil：ゴムは，鉱油で害を受けがちである（it = rubber）
   * mineral oil：〔名詞〕鉱油，鉱物油
   * the rubber mounts：ゴム製の取付台
   * have to be protected：〔受動態〕～されなければならない
   * against：〔前置詞〕～に対して。向かい合う力，双方から力がグッとかかるイメージがある。
   * by means of ～：〔熟語〕～の手段によって，～をすることによって
   * cover：〔名詞〕おおい，カバー，ふた
   * so：〔接続詞〕そこで
   * vibration damping：振動減衰
   * damage：〔名詞〕損害
   * mounting：〔名詞〕取付け
   * protect：〔他動詞〕～を保護する
   * dripping：〔名詞〕（機械の油の）しずく
   * splashing：〔名詞〕（はねかかった）飛沫

   ①②③④部分訳 _____

   ⑧⑨⑩⑪部分 _____

全文和訳

8) ②Brackets for the rubber mounts ①**are fitted**/ ⑤on the foundation plate/ ④(using steel or resin chocks.) ③/

   * bracket：〔名詞〕腕木，張り出し棚受け（L字型アーム）
   * be fitted on ～：〔受動態〕～にはめ込まれる
   * steel：〔名詞〕鋼鉄，はがね
   * chock：〔名詞〕チョック，くさび，輪止め
   * using：〔現在分詞〕用いた（the foundation plate を修飾する形容詞の働き）
   * fit：〔他動詞〕（部品などを）物にはめ込む，備え付ける
   * foundation plate：〔名詞〕基礎板
   * resin：〔名詞〕樹脂

全文和訳

9) ①Buffers,/ ②similar to the flexible mountings,/ ⑤**are fitted**/ ④at the sides and ends/ ③of the engine./

   * buffer：〔名詞〕緩衝器（装置）
   * are fitted：〔受動態〕備え付けられる
   * (being) similar to：〔形容詞〕～に似ている
   * at the sides and ends of ～：～の側面と端面に

全文和訳

10) ①The flexible mountings ④**must be** carefully ④**designed**/ ③to ensure the attenuation of structure borne noise./

   * flexible mounting：伸縮性のある取付け
   * must be carefully designed：〔受動態〕注意深く設計されなければいけない
   * attenuation：〔名詞〕（エネルギーなどの）弱化，減衰
   * structure borne noise：構造伝達騒音
   * ensure：〔他動詞〕確実に～をする，～を確認する，保証する

②③部分訳

全文和訳

Step 3　網掛けは主語，**ゴシック体**は動詞（述語動詞）

(1) 音声を聞きながら，音読する。音読が終わったら，Step Check Box □をぬりなさい。
(2) 左から順に，スラッシュ毎の意味を考える。

The various kinds of flexible or resilient mounting all **serve to keep** the engine/ in the desired position/ and **to prevent** transmission of vibration/ to the hull of the ship./ In passenger ships/ it **is** essential/ to avoid vibration and structure borne noise/ in the accommodation spaces./ but the use of medium speed engines **allows** passenger spaces/ to be placed low down/ in the ship./ It **is** possible/ to mount engines/ on a raft with integral vibration damping,/ but direct mounting of engines in the ship **is** simpler and less costly./ Flexible/resilient mountings **are** rubber based/ and **may be**/ of the inclined or vertical type./ The rubber pads **absorb** noise and vibration energy./ An effective resilient mounting system **will provide** an isolation factor/ of at least 50 or 80 per cent./ Although rubber **has** excellent vibration damping properties,/ it **is prone to** damage/ by mineral oil,/ so the rubber mounts **have to be protected**/ against dripping and splashing/ by means of covers./ Brackets for the rubber mounts **are fitted**/ on the foundation plate/ using steel or resin chocks./ Buffers,/ similar to the flexible mountings,/ **are fitted**/ at the sides and ends/ of the engine./ The flexible mountings **must be** carefully **designed**/ to ensure the attenuation of structure borne noise./

【専門英語単語リスト】書いておぼえよう

発音したり，スペル練習したり，意味を調べたりしたらチェック欄の○をぬりなさい。
辞書を引いて単語の意味が多くある場合，英語の文脈から判断し，適切な意味を選びなさい。

英文	単語	品詞	意味を調べて書きなさい	発音しながらスペル練習しなさい	チェック
1)	flexible	形容詞			○○○
1)	resilient	形容詞			○○○
1)	mounting / mount	名 / 動			○○○
1)	serve to V	自動詞			○○○
1)	prevent	他動詞			○○○
2)	accommodation	名詞			○○○
2)	allow ＋名詞＋ to V	他動詞			○○○
3)	damping	名詞			○○○
5)	absorb / absorption	動 / 名			○○○
6)	isolation / isolate	名 / 動			○○○
7)	rubber	名詞			○○○
7)	protect / protective	動 / 形			○○○
7)	by means of ～	熟語			○○○
8)	resin	名詞			○○○
9)	be fitted	受動態			○○○

理解を深めるキーワード「主機の据付け」

　エンジンの据付け（Mounting）において考慮すべき重要な要素は防振対策である。

　エンジンを起震源とする振動・騒音は機関の回転数が高速になるほど船体振動のみならず船内騒音への影響度が大きくなることが一般的であり，豪華さを売り物とする客船では船体構造（Structure）を経由（borne）して客室区画など（Accommodation space）に伝わる騒音はできうる限り避けなければならない。

　しかし，客船においては，客に対するサービススペースを大きく取るために機関室スペースを小さくすることが求められ，結果として中速機関（Medium speed engine）（背丈が小さく，容積に比べて出力が出る。回転数200～400rpm）がよく使われている。

　回転数の大きい中速機関では厳重な防振対策が必要となり，現状では防振に大きな効果があるとされるゴム製の防振装置（Absorber, Damper：下図参照）が多用されている。ゴム製は減衰効果が非常に大であるが，高温や油に浸漬すると膨潤して軟化し，破壊するという欠点を有している。よって，油の飛沫や滴下から防振装置を守ることが据付けや維持管理上たいへん重要である。

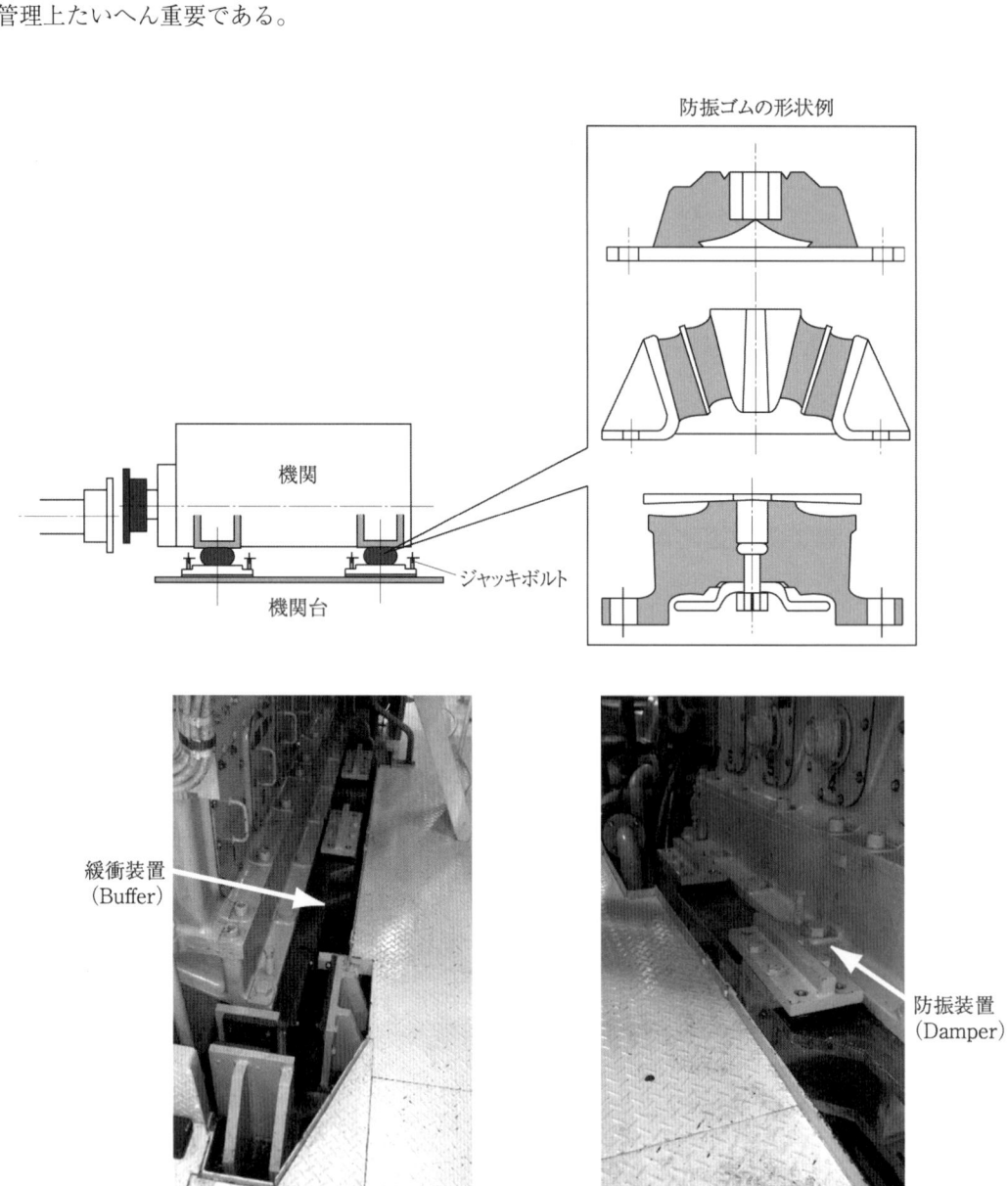

コラム5

前置詞3　熟語について

知らない単語は辞書で調べればいいけれど，熟語はどの部分が熟語なのかがわからないので辞書で調べるもの難しいと思っている人は，次のことを覚えておくと熟語が見つけやすくなります。

（前置詞＋名詞）の部分の意味が分からないときは

1．「前置詞」の前か後の単語を辞書でひいて「前置詞」と熟語になっていないか調べる。
2．「前置詞」より前の動詞（間にいくつか単語があっても可）と熟語になっていないか調べる。

これですべての熟語が見つけられるわけではありませんが，前置詞が含まれる熟語は多いので，熟語を見つけるのがずいぶん楽になるはずです。たとえば，この教科書には次のような熟語が出てきます。どの部分が熟語かわかりますか？　前置詞には下線を引いてあります。

… he is fully aware of the action ….
The procedure to replace it must be followed in accordance with the instruction manual to prevent damage.
This would result in increased impeller friction ….
Further, they both depend on the same electromagnetic principles for their operation.
… so the rubber mounts have to be protected against dripping and splashing by means of covers.

次の下線部が熟語で，太字の単語を辞書で引くと熟語が見つかります。
… he is fully **aware** of the action ….
The procedure to replace it must be followed in **accordance** with the instruction manual to prevent damage.
This would **result** in increased impeller friction ….
Further, they both **depend** on the same electromagnetic principles for their operation.
… so the rubber mounts have to be protected against dripping and splashing by **means** of covers.

Unit 38　高温腐食

Step Check Box □□□

Step 1　下記の英文を読み，次の３つの活動をしなさい。終了後 Step Check Box □をぬりなさい。

(1) 動詞（述語動詞）を○で囲む。動詞の意味を考えて，動詞のあとに続く内容を推測する。
(2) 主語（主部）に下線を引く。
(3) スラッシュを入れる。スラッシュを入れる場所は主に下記の３つ。
- コンマ(,)やピリオド(.)でスラッシュ(/)を入れる。音読するときは，スラッシュで息つぎする。
- 接続詞の前にスラッシュを入れる。［接続詞＋主語＋動詞］は文としてまとまった意味を表す。
  ここでいう接続詞とは when, if, until, before, after など（and, but, or, so を含まない）。
- 原則として前置詞の前にスラッシュを入れる。（前置詞＋名詞）は，最小の意味のまとまりを表す。

1) High-temperature corrosion
2) Vanadium is the major fuel constituent influencing high-temperature corrosion.
3) It cannot be removed in the pre-treatment process and it combines with sodium and sulphur during the combustion process to form eutectic compounds with melting points as low as 530℃.
4) Such molten compounds are very corrosive and attack the protective oxide layers on steel, exposing it to corrosion.
5) Exhaust valves and piston crowns are very susceptible to high-temperature corrosion.
6) One severe form is where mineral ash deposits form on valve seats, which, with constant pounding, cause dents leading to a small channel through which the hot gases can pass.
7) The compounds become heated and then attack the metal of the valve seat.
8) As well as their capacity for corrosion, vanadium, sulphur and sodium deposit out during combustion to foul the engine components and, being abrasive, lead to increased liner and ring wear.

（注）eutectic compounds：共晶化合物（共融混合物）

出典：１級海技士（機関）平成 24 年 4 月定期試験問題

Step 2　網掛けは主語，ゴシック体は動詞（述語動詞），＊語注・文法的留意事項，①②③は訳順番号

(1) 単語や熟語の説明を読みながら，部分訳を書きなさい。わからない単語の意味は調べなさい。
(2) 部分訳が終わったら，①②③の訳順番号を参考に，全文和訳を書きなさい。
　全文和訳が終わったら，Step Check Box □をぬりなさい。

1) High-temperature corrosion
　＊「高温腐食」がタイトルなので，下記には，タイトルに関する情報が書かれている。
　＊ corrosion：〔名詞〕腐食，さび　　　＊ corrosive：〔形容詞〕（酸・アルカリが）腐食性の，腐食する ⇔ noncorrosive

タイトル和訳

2) Vanadium is the major fuel constituent/ influencing high-temperature corrosion./
　　①　　　④　　　　　③　　　　　　　　　　　　　　　②

　＊ vanadium：〔名詞〕バナジウム（V）
　　モリブデン-バナジウム鋼，クロム-バナジウム鋼などの高張力鋼の製造に用いられる。
　＊ constituent：〔名詞〕構成物質（要素），成分　　＊ constitute：〔他動詞〕〜を構成する
　＊ fuel constituent：燃料物質
　＊ influencing：〔現在分詞〕名詞 constituent を修飾する。　＊ influence：〔他動詞〕〜に影響を及ぼす

全文和訳

Unit 38 高温腐食

3) It **cannot be removed**/ in the pre-treatment process/ and it **combines**/ with sodium and sulphur/ during the combustion process/ to form eutectic compounds/ with melting points/ as low as 530℃./

- ＊It：前の文の主語 vanadium を指す。
- ＊cannot be removed：〔受動態〕除去されない
- ＊cannot be removed in the pre-treatment process：前処理の段階で除去されない
- ＊combine with ～ to form …：〔熟語〕～と化合して，…を形成する
- ＊sodium and sulphur：ナトリウム (Na) と硫黄 (S)
- ＊sulphur：イギリス英語のスペル（アメリカ英語のスペルは sulfur）
- ＊during：〔前置詞〕（一定）の間
- ＊combustion：〔名詞〕燃焼
- ＊process：〔名詞〕過程，工程
- ＊eutectic：〔形容詞〕共晶（共融）混合物の
- ＊compound：〔名詞〕混合物，化合物
- ＊form：〔他動詞〕～を形成する
- ＊with melting point：融点で
- ＊as low as 530℃：530℃ ほどの低い
- ＊as 形容詞 as ～：同等比較を表す。

①②③部分訳＿＿＿＿＿＿＿＿＿＿＿＿＿＿＿＿＿＿＿＿＿＿＿＿＿＿＿＿＿＿＿＿＿

④⑤⑥⑦部分訳＿＿＿＿＿＿＿＿＿＿＿＿＿＿＿＿＿＿＿＿＿＿＿＿＿＿＿＿＿＿＿＿

⑧⑨⑩部分訳＿＿＿＿＿＿＿＿＿＿＿＿＿＿＿＿＿＿＿＿＿＿＿＿＿＿＿＿＿＿＿＿＿

全文和訳＿＿

4) Such molten compounds **are** very corrosive/ and **attack** the protective oxide layers on steel,/ exposing it to corrosion./

- ＊molten：〔形容詞〕（金属，岩石などが）熱で融けた，溶融した（状態の）
- ＊compound：〔名詞〕混合物
- ＊attack：〔他動詞〕～を腐食させる
- ＊protective oxide layer：保護酸化被膜
- ＊protective：〔形容詞〕防護用
- ＊oxide：〔名詞〕酸化物，酸化
- ＊layer：〔名詞〕層
- ＊on steel：鋼鉄上の
- ＊expose：〔他動詞〕さらす
- ＊exposing it (= steel) to corrosion：exposing は現在分詞。「そして鋼鉄を腐食させる」となる。exposing の前にカンマが付いているので，steel を修飾しない。「それから～になる」という意味を表す。〔接続詞＋主語 S ＋動詞 V → and they expose it to corrosion〕に書き変えることができる。

全文和訳＿＿

5) Exhaust valves and piston crowns **are** very susceptible/ to high-temperature corrosion./

- ＊exhaust valve：排気弁
- ＊piston crown：ピストンクラウン，ピストンヘッド
- ＊be susceptible to ～：～に影響されやすい，（作用）を受けやすい
- ＊high-temperature corrosion：高温腐食

全文和訳＿＿

6) One severe form **is** (a place) [where mineral ash deposits **form**/ on valve seats,]/

- ＊one severe form：1 つの重大な形状
- ＊severe：〔形容詞〕（病気などが）重い
- ＊mineral：〔名詞〕（動植物に対して）無機物
- ＊deposit：〔名詞〕堆積物，付着した燃焼生成物
- ＊ash：〔名詞〕燃えがら，灰
- ＊mineral ash deposit：無機質燃焼生成物
- ＊form：〔自動詞〕～が形を成す，発生する
- ＊valve seat：弁座
- ＊where：〔関係副詞〕先行詞 a place が省略されていて，where 以下が修飾する。
- ＊②③④⑤の大意：～は，弁座に無機質燃焼生成物が付着する（形成される）所である

①②③④⑤部分訳＿＿＿＿＿＿＿＿＿＿＿＿＿＿＿＿＿＿＿＿＿＿＿＿＿＿＿＿＿＿＿

163

　　　　⑥　　　　　　　⑦　　　　　　　　⑪　　⑩　　　　　　⑨　　　　　　　　　　⑧
[which,/ with constant pounding,/ **cause** dents/ leading to a small channel]/ [through which the hot gases **can pass**.]/

- pounding：〔動名詞〕（ハンマなどで）何度も強く打つこと
- constant：〔形容詞〕休みなく
- cause：〔他動詞〕～の原因となる
- dent：〔名詞〕へこみ，くぼみ
- leading to a small channel：leading は現在分詞で dents にかかる。
- small channel：小さい溝
- which (= valve seats) cause dents ～：which は関係代名詞で，valve seats を指す。which の前にカンマが付いているので，and valve seats cause dents leading to a small channel と書き換えることができる。
- ⑥⑦⑧⑨⑩の大意：そして，（鉱物性の灰堆積物が付着している）弁座は，絶えず叩かれることによって，⑧のような小さい溝となるへこみの原因となる
- through which：which 以下は a small channel にかかる。
- ⑧の大意：熱ガスが通過するような（小さい溝）

全文和訳

　　　　　　　①　　　　　②　　　　　　　　⑤　　　④　　　　　③
7) The compounds **become** heated/ and then **attack** the metal/ of the valve seat./

- become heated：熱せられる

全文和訳

　　　　　　　　　　　　①　　　　　　　　　　　　②　　　　　　　　④　　　　③
8) As well as their capacity for corrosion,/ vanadium, sulphur and sodium **deposit out**/ during combustion/
　　　⑤　　　　　　　　　⑥　　　　　⑧　　　　　⑦
to foul the engine components/ and, (being abrasive),/ **lead to** increased liner and ring wear./

- as well as：〔前置詞〕～に加えて
- capacity：〔名詞〕容量，容積，包含する能力
- as well as their capacity for corrosion：（化合物の）腐食性に加えて
- deposit out：〔自動詞〕～が積もる，堆積する
- the engine components：機関部品
- combustion：（化学）燃焼，（有機体の）酸化
- foul：〔他動詞〕～を（異物で）覆う，～を汚す
- to foul：不定詞（to + 動詞）として働き，「結果として～して」という意味を表す。
- ②③④⑤の大意：燃焼の間，②が堆積して，結果として機関部品をよごす
- being abrasive：being は現在分詞。because they are abrasive と同じ意味。「（堆積物で）研磨されるので」となる。
- abrasive：〔形容詞〕すり減らす（作用をする）
- increased liner and ring wear：ライナ（かぶせ金）とリングの摩耗の増大
- lead to：（物事がある結果に）つながる，引き起こす

全文和訳

Step 3　網掛け は主語，ゴシック体は動詞（述語動詞）

(1) 音声を聞きながら，音読する。音読が終わったら，Step Check Box □ をぬりなさい。
(2) 左から順に，スラッシュ毎の意味を考える。

High-temperature corrosion
　Vanadium **is** the major fuel constituent/ influencing high-temperature corrosion./ It **cannot be removed**/ in the pre-treatment process/ and it **combines**/ with sodium and sulphur/ during the combustion process/ to form eutectic compounds/ with melting points/ as low as 530℃./ Such molten compounds **are** very corrosive/ and **attack** the protective oxide layers on steel,/ exposing it to corrosion./
　Exhaust valves and piston crowns **are** very susceptible/ to high-temperature corrosion./ One severe form **is** where mineral ash deposits **form**/ on valve seats,/ which,/ with constant pounding,/ **cause** dents/ leading to a

164

small channel/ through which the hot gases can pass./ The compounds become heated/ and then attack the metal/ of the valve seat./

As well as their capacity for corrosion,/ vanadium, sulphur and sodium deposit out/ during combustion/ to foul the engine components/ and, being abrasive,/ lead to increased liner and ring wear./

【専門英語単語リスト】書いておぼえよう

英文	単語	品詞	意味を調べて書きなさい	発音しながらスペル練習しなさい	チェック
1)	corrosion / corrosive	名／形			○○○
2)	constituent / constitute	名／動			○○○
2)	influence	他動詞			○○○
3)	combine / combination	動／名			○○○
4)	compound	名詞			○○○
4)	attack	他動詞			○○○
4)	expose ～ to …	他動詞			○○○
5)	susceptible	形容詞			○○○
6)	mineral ash deposit	名詞			○○○
6)	form	自動詞			○○○
6)	dent	名詞			○○○
8)	as well as	熟語			○○○
8)	deposit out	自動詞			○○○
8)	combustion / combust	名／動			○○○
8)	foul	他動詞			○○○

理解を深めるキーワード「高温腐食とバナジウムアタック」

重油燃料中に含まれるバナジウム (V, vanadium) は燃焼により酸化し，燃料中のナトリウム (Na, sodium)，Ca, Ni, Fe と反応してバナジウム化合物を形成する。このような化合物が高温状態の金属表面に接触すると，その表面温度が融点以上であれば融解する（化合物 $2Na_2O \cdot 3V_2O_5$ の融点は 530℃ 程度）。その際に金属表面の酸化皮膜を破壊して腐食が進むことから，「高温腐食 (High-temperature corrosion)」および，バナジウムに起因するために「バナジウムアタック (Vanadium attack)」と呼ばれる。また，このようなバナジウム化合物の融解による酸化腐食において，燃料中の硫黄分 (S, sulphur, sulfur) も酸化を促進させる成分となる。

たとえば

$$Na_2 \cdot 6V_2O_5 + Fe \rightarrow FeO + Na_2 \cdot V_2O_4 \cdot 5V_2O_5$$

により，ナトリウム，酸素と結合したバナジウム化合物 ($Na_2 \cdot 6V_2O_5$) が鉄 (Fe) と反応して酸化鉄 (FeO) が形成され，形を変えたナトリウム，酸素，バナジウムの化合物 ($Na_2 \cdot V_2O_4 \cdot 5V_2O_5$) が残る。

CHAPTER 4 の Unit 52 では，実際にピストンクラウン表面で生じたバナジウムアタックによる腐食がピストン定期検査で発見されたことが記されている。

Unit 39　ピストン

Step Check Box □□□

Step 1　下記の英文を読み，次の3つの活動をしなさい。終了後 Step Check Box □ をぬりなさい。

(1) 動詞（述語動詞）を○で囲む。動詞の意味を考えて，動詞のあとに続く内容を推測する。
(2) 主語（主部）に下線を引く。
(3) スラッシュを入れる。スラッシュを入れる場所は主に下記の3つ。
- コンマ (,) やピリオド (.) でスラッシュ (/) を入れる。音読するときは，スラッシュで息つぎする。
- 接続詞の前にスラッシュを入れる。[接続詞＋主語＋動詞] は文としてまとまった意味を表す。
  ここでいう接続詞とは when, if, until, before, after など (and, but, or, so を含まない)。
- 原則として前置詞の前にスラッシュを入れる。(前置詞＋名詞) は，最小の意味のまとまりを表す。

1) Piston crowns can be visually inspected from above by lifting the cylinder head, but a thorough inspection requires the piston to be lifted.
2) The connecting rod, or a section of it, will be lifted with the piston.
3) The correct lifting tool must be used to avoid damage to the crown and, in the case of Vee type engines, to ensure that the piston is pulled cleanly up the cylinder bore.
4) All threaded lifting holes in the piston crown should be cleaned before attaching the lifting tool.
5) Routine piston lifting should take place at intervals of about 8000 hours to 15000 hours, when piston ring packs are replaced.
6) The gudgeon pin assembly should be checked for security and freedom but should not be dismantled until about 24000 running hours have elapsed unless defects are detected at an earlier examination.
7) With floating gudgeon pins the procedure should require no force, but a fitted pin will need to be driven out of the piston after the retaining clips have been removed.
8) Gudgeon pin bearing surfaces should be checked for signs of wear of cracking, and the bores measured.

（注）Vee type engines：V型機関

出典：1級海技士（機関）平成24年4月定期試験問題

Step 2　網掛け は主語，**ゴシック体**は動詞（述語動詞），＊語注・文法的留意事項，①②③は訳順番号

(1) 単語や熟語の説明を読みながら，部分訳を書きなさい。わからない単語の意味は調べなさい。
(2) 部分訳が終わったら，①②③の訳順番号を参考に，全文和訳を書きなさい。
　全文和訳が終わったら，Step Check Box □ をぬりなさい。

1) ①Piston crowns ④**can be** visually **inspected**/ ③from above/ ②by lifting the cylinder head,/ but ⑤a thorough inspection ⑧**requires** ⑥the piston ⑦to be lifted./

- ＊ piston crown：〔名詞〕ピストンクラウン，ピストンヘッド
  ピストン：エンジンのシリンダ内を往復運動する円筒状の部分
- ＊ can be visually inspected：〔受動態〕視覚的に点検される　＊ visually：〔副詞〕視覚的に
- ＊ inspect：〔他動詞〕～を点検する　＊ inspection：〔名詞〕点検，調査，視察
- ＊ from above：上部から　＊ by lifting ～：～することにより（lifting は動名詞）
- ＊ lift：〔他動詞〕～を持ち上げる，手にとって下ろす，～を持ち上げて出す
- ＊ require A to 不定詞：A が～されるよう求める　＊ to be lifted：〔不定詞の受動態〕持ち上げて出される
- ＊ a thorough inspection：完全な点検
- ＊ この英文の主語は piston crown で，動詞は can be inspected であるから，トピックがピストンの点検のことだとわかる。試験で問われているトピックが，1つの段落（パラグラフ）のなかで展開されることを理解しておく。

⑤⑥⑦⑧部分訳　しかし，完璧な点検（の場合）は _____

全文和訳
_____

2) The connecting rod,/ or a section of it,/ **will be lifted**/ with the piston./
   　　①　　　　　　　　②　　　　　　④　　　　　③

* connecting rod：コンロッド，連接棒　　　* section：〔名詞〕断面，薄片
* with：〔前置詞〕～といっしょに

全文和訳

3) The correct lifting tool **must be used**/ to avoid damage to the crown/ and, in the case of Vee type engines,/
   　　　①　　　　　　　③　　　　　　　　②　　　　　　　　　　　　　　　　　　④
   (the correct lifting tool **must be used**) to ensure [that the piston **is pulled** cleanly up the cylinder bore.]/
   　　　　　　　　　　　　　　　　　　⑥　　　　　⑤

* correct lifting tool：正規の吊り上げ工具
* to avoid damage to ～：～への損傷を避けるために（不定詞の用法「～するために」）
* in the case of 名詞：～の場合には　　　* Vee type engines：V型機関
* to ensure that 主語 S ＋動詞 V：to ensure は不定詞。「～ that 以下を確認するために」となる。
* cylinder bore：シリンダ内径（穴，口径）　* be pulled up = be lifted
* cleanly：〔副詞〕なめらかに，均一に
* ⑤⑥の大意：確実にピストンがシリンダ内径をなめらかに持ち上げられるために（①が使用されねばならない）

①②③部分訳＿＿＿＿＿＿＿＿＿＿＿＿＿＿＿＿＿＿＿＿＿＿＿＿＿＿＿＿＿＿＿

全文和訳

4) All threaded lifting holes/ in the piston crown **should be cleaned**/ before attaching the lifting tool./
   　　　　②　　　　　　　　①　　　　　　　④　　　　　　　　　　　③

* all threaded lifting holes：すべてのネジ筋（山）の吊り上げ用穴
* attach：〔他動詞〕～を取り付ける，はり付ける　* the lifting tool：吊り上げ工具

全文和訳

5) Routine piston lifting **should take place**/ at intervals of about 8000 hours to 15000 hours,/
   　　①　　　　　　　⑥　　　　　　　　　　　　　　⑤
   [when piston ring packs **are replaced**.]/
   　④　　　②　　　　　　　③

* routine piston lifting：定期的なピストン吊り上げ　* take place：〔熟語〕行われる
* at intervals of：～間隔で　　　　　　　　* replace：〔他動詞〕取り替える
* piston ring pack：ピストンリング一式（一包み）　* be replaced：〔受動態〕取り替えられる

全文和訳

6) The gudgeon pin assembly **should be checked**/ for security and freedom/ but (the gudgeon pin assembly)
   　　　①　　　　　　　　③　　　　　　　　②　　　　　　　　　　　⑪
   should not be dismantled/ [until about 24000 running hours **have elapsed**] [unless defects **are detected**/
   　⑫　　　　　⑩　　　　　　　　⑧　　　　　　　　　⑨　　　　　⑦　　⑤　　　　⑥
   at an earlier examination.]/
   　　　④

* gudgeon pin assembly：ガジオンピン一式，ピストンピン一式　* assembly：〔名詞〕組み立て部品一式
* for security and freedom：安全と故障なし（freedom）のためには
* should not be dismantled：分解されるべきではない
* unless：〔接続詞〕（= if not）もし～でないならば　　*接続詞が否定語の場合は，動詞に not を付けて訳す。
* defect：〔名詞〕欠陥，きず（= fault）　　　* at an earlier examination：初期の検査で
* be detected：検出される，の存在が発見される

④⑤⑥⑦部分訳＿＿＿＿＿＿＿＿＿＿＿＿＿＿＿＿＿＿＿＿＿＿＿＿＿＿＿＿＿＿＿

* until：〔接続詞〕〜するまで（接続詞の後には，主語 S ＋動詞 V が続く）
* running hour：運転時間
* elapse：〔自動詞〕時が経つ，経過する
* until 〜 have elapsed：経過してしまうまで
  until 以下は未来の内容を指すが，未来形を使わない。
  時や条件を表す副詞節（when, until, if, unless, before, after など＋主語 S ＋動詞 V）のなかでは，未来のことでも現在形（or 現在完了形）で表現する。

⑧⑨⑩部分訳＿＿＿＿＿＿＿＿＿＿＿＿＿＿＿＿＿＿＿＿＿＿＿＿＿＿＿＿

全文和訳 ＿＿＿

7) With floating gudgeon pins/ ① the procedure ② should require ④ no force,/ but ③ a fitted pin ⑤ will need ⑨ to be driven/ ⑪ out of the piston/ ⑩ [after ⑧ the retaining clips ⑥ have been removed.]/ ⑦

* with floating gudgeon pins：浮動式ピストンピンがあると
* procedure：〔名詞〕手順
* no force：労力を〜しない（名詞に否定語 no が付いているときは，動詞を否定するように訳す）
* a fitted pin：固定式ピン
* will need to be driven：抜き出す（される）必要がある
* retaining clip：保持金具
* have been removed：現在完了形の受動態
* remove：〔他動詞〕取り除く
* after 〜 have been removed：after（接続詞）以下は未来のことでも現在形（現在完了形）で書く。
  （接続詞＋主語 S ＋動詞 V）は，動詞（will need）にかかる（修飾する）。

②③④部分訳＿＿＿＿＿＿＿＿＿＿＿＿＿＿＿＿＿＿＿＿＿＿＿＿＿＿

⑨⑩⑪部分訳 （しかし）＿＿＿＿＿＿＿＿＿＿＿＿＿＿＿＿＿＿＿＿＿＿＿＿＿＿

⑥⑦⑧部分訳＿＿＿＿＿＿＿＿＿＿＿＿＿＿＿＿＿＿＿＿＿＿＿＿＿＿の後

全文和訳 ＿＿＿

8) ① Gudgeon pin bearing surfaces should be checked/ ③ for signs of wear of cracking,/ ② and ④ the bores (should be) measured./ ⑤

* gudgeon pin bearing surface：ガジオンピン軸受面（表面）
* for signs of wear of cracking：摩耗して亀裂する前兆がないか
* bore：〔名詞〕内径
* (should be) measured：計測されるべきである

全文和訳 ＿＿＿

Step 3   網掛け は主語，**ゴシック体**は動詞（述語動詞）

(1) 音声を聞きながら，音読する。音読が終わったら，Step Check Box □をぬりなさい。
(2) 左から順に，スラッシュ毎の意味を考える。

Piston crowns **can be** visually **inspected**/ from above/ by lifting the cylinder head,/ but a thorough inspection **requires** the piston to be lifted./ The connecting rod,/ or a section of it,/ **will be lifted**/ with the piston./ The correct lifting tool **must be used**/ to avoid damage to the crown/ and, in the case of Vee type engines,/ to ensure that the piston **is pulled** cleanly **up** the cylinder bore./ All threaded lifting holes/ in the piston crown **should be cleaned**/ before attaching the lifting tool./ Routine piston lifting **should take place**/ at intervals of about 8000 hours to 15000 hours,/ when piston ring packs **are replaced**./ The gudgeon pin assembly **should be checked**/ for security and freedom/ but **should not be dismantled**/ until about 24000 running hours **have elapsed**/ unless defects **are detected**/ at an earlier examination./ With floating gudgeon pins/ the procedure **should require** no force,/ but a fitted pin **will need** to be driven/ out of the piston/ after the retaining clips **have been removed**./ Gudgeon pin bearing surfaces **should be checked**/ for signs of wear of cracking,/ and the bores **measured**./

【専門英語単語リスト】書いておぼえよう

発音したり，スペル練習したり，意味を調べたりしたらチェック欄の〇をぬりなさい。
辞書を引いて単語の意味が多くある場合，英語の文脈から判断し，適切な意味を選びなさい。

英文	単語	品詞	意味を調べて書きなさい	発音しながらスペル練習しなさい	チェック
1)	inspect / inspection	動 / 名			〇〇〇
1)	thorough	形容詞			〇〇〇
1)	require / requirement	動 / 名			〇〇〇
2)	connecting rod	名詞			〇〇〇
3)	cylinder bore	名詞			〇〇〇
4)	attach	他動詞			〇〇〇
5)	take place	自動詞			〇〇〇
6)	gudgeon pin	名詞			〇〇〇
6)	dismantle	他動詞			〇〇〇
6)	until / unless	接続詞			〇〇〇
6)	defect / defective	名 / 形			〇〇〇
6)	detect	他動詞			〇〇〇
7)	procedure	名詞			〇〇〇
7)	drive	他動詞			〇〇〇
7)	retain	他動詞			〇〇〇

〜 理解を深めるキーワード「ピストン」 〜

　ピストンは，頂部での燃焼による燃焼圧を受けて往復運動し，連接棒を介してクランク軸に回転力を与える。燃焼による高温・高圧の状況にさらされるため，その材質としては耐熱・耐圧性のもので耐久力の大きいこと，比較的熱伝導率が高く，軽いことが望まれる。ピストンはピストンと連接棒がピストンピン (Piston pin, Gudgeon pin) でつながれた構造となっており，ピストン上部のことをとくにピストンクラウン (Piston crown)，下部をスカートと呼ぶ。ピストンピンの止め方により浮動式 (Floating gudgeon pin) と固定式 (Fitted pin) がある。

　本問題では高温・高圧下で運動するピストンの検査ための手順とその検査を行う運転時間の目安，ピストンピンやその軸受の検査方法について記載されている。また，ピストンの構造の図やピストンクラウンで生じた焼損についてはCHAPTER 4のUnit 53に記されている。

Unit 40　スラッジ処理　　Step Check Box □ □ □

Step 1　下記の英文を読み，次の3つの活動をしなさい。終了後 Step Check Box □をぬりなさい。

(1) 動詞（述語動詞）を○で囲む。動詞の意味を考えて，動詞のあとに続く内容を推測する。
(2) 主語（主部）に下線を引く。
(3) スラッシュを入れる。スラッシュを入れる場所は主に下記の3つ。
- コンマ (,) やピリオド (.) でスラッシュ (/) を入れる。音読するときは，スラッシュで息つぎする。
- 接続詞の前にスラッシュを入れる。［接続詞＋主語＋動詞］は文としてまとまった意味を表す。
  ここでいう接続詞とは when, if, until, before, after など（and, but, or, so を含まない）。
- 原則として前置詞の前にスラッシュを入れる。（前置詞＋名詞）は，最小の意味のまとまりを表す。

1) The problems of storage in tanks of bunker fuel result from a build-up in sludge leading to difficulties in handling.
2) The reason for the increase in sludge build-up is because heavy fuels are generally blended from a cracked heavy residual using a lighter cutter stock resulting in a problem of incompatibility.
3) This occurs when the asphaltene or high molecular weight compound suspended in the fuel is precipitated by the addition of the cutter stock or other dilutents.
4) The sludge which settles in the bunker tanks or finds its way to the fuel lines tends to overload the fuel separators with a resultant loss of burnable fuel, and perhaps problems with fuel injectors and wear of the engine through abrasive particles.
5) To minimize the problems of sludging the ship operator has a number of options.
6) He may ask the fuel supplier to perform stability checks on the fuel that he is providing.
7) Bunkers of different origins should be kept segregated wherever possible and water contamination kept to a minimum.

（注）cutter stock：カッター材　dilutents = diluents

出典：1級海技士（機関）平成24年2月定期試験問題

Step 2　網掛けは主語，ゴシック体は動詞（述語動詞），＊語注・文法的留意事項，①②③は訳順番号

(1) 単語や熟語の説明を読みながら，部分訳を書きなさい。わからない単語の意味は調べなさい。
(2) 部分訳が終わったら，①②③の訳順番号を参考に，全文和訳を書きなさい。
　全文和訳が終わったら，Step Check Box □をぬりなさい。

1) The problems of storage/ in tanks of bunker fuel **result from** a build-up in sludge/ leading to difficulties in handling./
　　②　　　　　　　　　　①　　　　　　　　　⑤　　　　　　　④　　　　　　　　　③

- ＊ storage：〔名詞〕貯蔵
- ＊ bunker：〔名詞〕（主に船の）燃料
- ＊ bunker fuel：バンカー重油，船舶用燃料油
- ＊ fuel：〔名詞〕燃料
- ＊ result from：〔動詞〕〜から結果として生じる，〜に起因する
- ＊ build-up：〔名詞〕集積，蓄積
- ＊ sludge：（タンクなどのなかの）沈積物，スラッジ
- ＊ leading to 〜：〔現在分詞〕〜にいたる，〜に引き入れる (sludge にかかる)
- ＊ difficulties in handling：処理困難（処理するのが困難）
- ＊ ③④⑤の大意：処理困難になるスラッジの集積に起因する

全文和訳

2) The reason for the increase in sludge build-up **is**/ [because heavy fuels **are** generally **blended**/ from a cracked heavy residual/ (by) using a lighter cutter stock/ resulting in a problem of incompatibility.]/
　　　　　①　　　　　　　　　　　　　　　　　　　　④　　　　　　⑥　　　　　　⑤　　　　　　　　③　　　　　　　　②

- * increase：〔名詞〕増加（increase に the が付いているので，品詞は名詞）
- * sludge build-up：スラッジの集積
- * 主語 S + is [because　主語 S + 動詞 V]：The reason（理由）は，because 以下…だからである
- * ①の大意：スラッジ集積が増加する理由は
- * heavy fuel：〔名詞〕重油
- * residual：〔名詞〕残留物
- * heavy：〔形容詞〕大量の
- * a lighter cutter stock：より軽いカッター材
  cutter stock（カッター材）：減圧蒸留残油や PDA（プロパン脱瀝）アスファルトは，そのままでは高粘度のため燃料油として取扱いが困難である。そのため，直留軽油，分解軽油などの低粘度油を混合して粘度を下げている。この粘度を下げるための軽油などをいう。
- * incompatibility：〔名詞〕不親和性
- * using a lighter cutter stock resulting in a problem of incompatibility：結果として不親和性の問題となる軽油を使うことによって（＝軽油を使用することで不親和性が問題となる）
- * ④⑤⑥の大意：重油が一般的に多量の分解された残渣に混合されてしまうからである
- * be blended from：〔受動態〕from 以下に混ぜ合わせられる
- * crack：〔他動詞〕（重油などを）熱分解する
- * result in ～：〔熟語〕～に帰着する，～に終わる

①部分訳 _____　　②③部分訳 _____

④⑤⑥部分訳 _____

全文和訳 _____

3) This **occurs**/ [when the asphaltene or high molecular weight compound/ suspended in the fuel **is precipitated**/ by the addition of the cutter stock or other dilutents.]/

- * asphaltene：アスファルテン（アスファルトの石油エーテル不溶性成分のうちベンゼン二硫化炭素に不溶性の高分子物質）
- * high molecular weight compound：高分子化合物
- * suspended：〔過去分詞〕懸濁する（compound にかかる）
- * precipitate：〔他動詞〕～を沈殿させる
- * addition：〔名詞〕追加
- * dilutent（＝ diluent）：〔名詞〕希釈剤，薄め液
- * be precipitated by ～：〔受動態〕～によって沈殿する
- * add：〔動詞〕加える
- * the asphaltene or high molecular weight compound suspended in the fuel：アスファルテンまたは燃料のなかで懸濁している高分子化合物
- * by the addition of the cutter stock or other dilutents：軽油または他の希釈剤を追加することによって

全文和訳 _____

4) The sludge [which **settles** in the bunker tanks/ or **finds** its way to the fuel lines]/ **tends to** overload the fuel separators/ with a resultant loss of burnable fuel,/ and (with) perhaps problems/ with fuel injectors and wear of the engine/ through abrasive particles./

- * settle：〔自動詞〕～が沈殿する
- * find its way to ～：〔熟語〕～にたどりつく
- * tend to overload：～に負荷をかけすぎる傾向にある
- * with resultant loss of：結果的に～の損失を伴い
- * resultant：〔形容詞〕結果として生じる，結果の
- * fuel injector：燃料噴射器（内燃機関の燃焼室に燃料を供給する装置）
- * wear of the engine：エンジンの摩損
- * through：〔前置詞〕（原因，理由，手段）～により，～のために
- * through abrasive particle：研磨粒子による
- * ⑤⑥⑦⑧の大意：おそらく燃料噴射器と研磨粒子によるエンジンの摩損という問題で
- * with：〔前置詞〕（原因，理由）が原因で，のために
- * with a resultant loss of burnable fuel and (with) perhaps problems：and は④と⑧以下をつなぎ，動詞 tend to にかかる
- * bunker tank：〔名詞〕燃料タンク
- * fuel line：〔名詞〕燃料管
- * fuel separator：〔名詞〕燃料分離器
- * burnable fuel：可燃性燃料
- * loss：〔名詞〕損失→ lose：〔他動詞〕～を失う
- * wear：〔名詞〕摩損，消耗

①②③部分訳 _____　　④部分訳 _____

⑤⑥⑦⑧部分訳 _____

全文和訳

5) To minimize the problems of sludging/ the ship operator **has** a number of options./

- * sludging：〔名詞〕スラッジが堆積すること
- * minimize：〔他動詞〕最小限度にする ↔ maximize
- * to minimize：〔不定詞〕～するために（動詞 has にかかる）
- * a number of ～：〔熟語〕数多くの～
- * ship operator：操船者
- * option：〔名詞〕選択肢

全文和訳

6) He **may ask** the fuel supplier/ to perform stability checks/ on the fuel [that he **is providing**.]/

- * ask 人 to 不定詞（to perform）：人に（が）～するように頼む
- * he：〔代名詞〕the ship operator を指す
- * may：〔助動詞〕（法律）…できる，…ねばならない = shall, must
- * the fuel supplier：燃料製造業者
- * that he is providing：本人（he = the fuel supplier）が供給している（that は関係代名詞，直前の名詞 fuel にかかる）
- * stability：〔名詞〕安定性
- * check on ～：～についての検査

全文和訳

7) Bunkers of different origins **should be kept** segregated/ [wherever (it is) possible]/ and water contamination (**should be**) **kept** to a minimum./]

- * different origins：産油地の違う燃料
- * bunker：〔名詞〕(船内の)燃料庫
- * wherever (it is) possible：(慣用的表現)可能なところではどこでも (it is が省略されている)
- * segregated：〔形容詞〕分離して
- * keep ＋目的語＋形容詞：～を…に保つ→（受動態にすると）should be kept ＋形容詞
- * water contamination：水濁
- * to a minimum：〔熟語〕最小限に

全文和訳

Step 3   網掛け は主語，**ゴシック体**は動詞（述語動詞）

(1) 音声を聞きながら，音読する。音読が終わったら，Step Check Box □をぬりなさい。
(2) 左から順に，スラッシュ毎の意味を考える。

The problems of storage/ in tanks of bunker fuel **result from** a build-up in sludge/ leading to difficulties in handling./ The reason for the increase in sludge build-up **is**/ because heavy fuels **are** generally **blended**/ from a cracked heavy residual/ using a lighter cutter stock/ resulting in a problem of incompatibility./ This **occurs**/ when the asphaltene or high molecular weight compound/ suspended in the fuel **is precipitated**/ by the addition of the cutter stock or other dilutents./ The sludge which **settles** in the bunker tanks/ or **finds** its way to the fuel lines/ **tends to** overload the fuel separators/ with a resultant loss of burnable fuel, and perhaps problems/ with fuel injectors and wear of the engine/ through abrasive particles./

To minimize the problems of sludging/ the ship operator **has** a number of options./ He **may ask** the fuel supplier/ to perform stability checks/ on the fuel that he **is providing**./ Bunkers of different origins **should be kept** segregated/ wherever possible/ and water contamination **kept** to a minimum./

【専門英語単語リスト】書いておぼえよう

発音したり，スペル練習したり，意味を調べたりしたらチェック欄の○をぬりなさい。
辞書を引いて単語の意味が多くある場合，英語の文脈から判断し，適切な意味を選びなさい。

英文	単語	品詞	意味を調べて書きなさい	発音しながらスペル練習しなさい	チェック
1)	storage / store	名 / 動			○○○
1)	bunker fuel	名詞			○○○
1)	result from ～	自動詞			○○○
1)	sludge	名詞			○○○
1)	lead to 名詞	自動詞			○○○
2)	build-up	名詞			○○○
2)	be blended from	受動態			○○○
2)	result in ～	自動詞			○○○
3)	occur / occurrence	動 / 名			○○○
3)	suspend	他動詞			○○○
3)	precipitate	他動詞			○○○
4)	overload	他動詞			○○○
5)	minimize / minimum	動 / 名			○○○
6)	supplier / supply	名 / 動			○○○
7)	contamination	名詞			○○○

コラム 6

接続詞 and の用法について

次のような形で箇条書きになった文が出てくることがあります。その最終行の手前に and という接続詞を見かけることがあります。

(a) _____ ;
(b) _____ ;
(c) _____ ; and
(d) _____ .

何も付いていない方がすっきりするはずなのに，なぜここに and があるのでしょうか。
接続詞の and は A and B のように2つのものをくっつけるだけでなく，A, B, and C のように3つ以上のものを並列する場合にも使われます。箇条書きのスタイルを見ても，今回の and は後者の用法になります。つまり，(a) から (d) までは並列していて，次の (d) が最後の部分であるというサインだと考えればいいわけです。その証拠に (a) ～ (c) の最後はコンマであったり，セミコロンであることが多いのですが，最後の (d) はちゃんとピリオドになっていることに気づくはずです。

コラム7

that の用法について

that を最初に覚えたのは,「あれ」とか「あの」という意味でしたよね。海技士の試験でいちばん多く出されるのは,「あれ」とか「あの」ではありません。逆に,「あれ」とか「あの」という意味はほとんど登場しません。ここでは,試験に出そうな使い方を整理しておきましょう。

① 接続詞の that
　訳し方は「～（という）こと」でいいでしょう。出る頻度が最も多いのがこの that です。
　The master shall ensure that ship's personnel always monitor cargo operations.
　（船長は船の乗組員がつねに荷役の作業を監視していることを確認しなければならない。）

② 関係代名詞の that
　2つの文をつなげる役割があります。うまくつなぐことができれば, that を訳す必要はありません。
　A ship that is carrying explosives flies a red flag.
　（爆発物を運んでいる船は赤い旗を掲げている。）

③ so ～ that…の構文
　「とても～なので…」と訳します。
　The ropes were so wet that they were left on the deck to dry.
　（ロープはかなり濡れていたので,乾かすために甲板上に置いておかれた。）

④ 同格の that．
　接続詞なのですが,①との違いは that の前に名詞があるということです。「～という…」と訳して,後ろの方から前にある名詞にかかっていきます。
　The reason wear is greatest here is due to the fact that friction is the greatest here.
　（ここの摩耗がいちばん大きい理由は,ここの摩擦が最も大きいという事実による。）

⑤ 指示代名詞の that
　「あれ」にいちばん近いのですが,少し使い方が凝っています。
　The climate of Japan is milder than that of England.
　（日本の気候はイギリス（の気候）よりも穏やかです。）（that = the climate）
　日本語では「の気候」と付けなくても通じますが,英語では比べるものが同等のものでなければならないので, that of が必要になるのです。

CHAPTER 3

専門英語 航海

Unit 41 　係留索の取扱い

Handling of Mooring Line

Mooring/unmooring work is one of most dangerous jobs for deck crew/ and sometimes fatal accidents happen to a seaman and/or a ship during the work./ We should keep it in mind/ that;

- **Do not stand** or **stay**/ where a rope **might bound back** or **rebound**/ in case of a rope whipping or breaking off suddenly./ **Avoid** straddling mooring ropes during mooring./
- At least 2 men **are to** be assigned to work together/ when using mooring winch and a rope./
- Signaling between a person in charge and workers **is** most important/ when preparing mooring ropes/ before entering or leaving a port./
- When slackening a mooring rope/ **walk** it back on a winch rather than releasing manually./
- **Maintain** a proper distance from the warping end of the mooring winch/ to avoid loose clothing being caught./
- When preparing hawsers on deck/ **beware** that if too much **is laid out**,/ it **can snake away** and **becomes** uncontrollable/ due to its weight./
- When securing a mooring rope on a winch drum,/ it **shall be wound** politely/ from one side toward the other side,/ not being jammed each other./
- A synthetic fiber hawser **elongates** and **reduces** in diameter under load./ It **is** most dangerous/ when stranding and recoiling./ You **should be** aware and **stand** well clear./
- A synthetic fiber hawser **deteriorates**/ when exposed to ultra violet rays./ They **should be covered**/ when not in use/ with a canvas cover, etc./
- Any protrusions or flaws on the fairleads or mooring barrels **may damage** synthetic fibers./ Rust **should be removed**,/ also grooving from wire ropes **are** also **damaging**./
- Chemical agent paints and thinners **will cause** damage/ to synthetic fiber hawsers./ They **should be stowed** well clear of these substances./

It **is** most dangerous/ when tension of the mooring rope **is transferred** to the stopper./ **Walk back** slowly and carefully/ until the stopper **holds** the tension firmly./ Signaling **is** most important./

- When securing a synthetic fiber hawser to a bollard,/ initial turn **should be** around the fore piece of the bollards/ and then **belay** in the shape of figure 8/ with at least 6 turns./ This method **allows** for easy adjustment to hawser length/ and **prevents** slipping under sudden weight./
- A chain stopper **should not be used** with synthetic hawsers./
- Double braided rope with sufficient strength **is recommended** for hawser stoppers/ to ensure that risk of snapping **is reduced**./
- Each side of a rope stopper **holds** the other end of the rope alternately/ at least three times./ Deck ring **is** sometimes not enough strong/ for the tension of mooring stopper./ In such case, a bollard **is to** be used as shown./

《 語　注 》

1. a person in charge：責任者
2. politely：丁寧に
3. elongate：引き伸ばす
4. deteriorate：劣化する
5. ultra violet ray：紫外線
6. protrusion：突出
7. flaw：傷
8. grooving：溝状腐食
9. firmly：しっかりと
10. double braided rope：二重組み紐

Reference：
International Mariners Management Association of Japan (IMMAJ), 2008, THE BEST SEAMANSHIP ―A Guide to Deck Skills―, Chapter 4：MOORING & UNMOORING

Unit 42　操縦性能の基礎

Fundamental Maneuvering Characteristics

Actual ship maneuvering patterns/ practiced under various navigational environments **are classified** broadly/ into two categories—course keeping and evasive (emergency) maneuvers./ When considering maneuvering procedures, such as course keeping, course changing, and decelerating/stopping, the following maneuvering characteristics **are required** for ship handling./

1. Turning ability

   Turning ability **is** the measure of the ability to turn the ship/ using hard-over rudder,/ the results being a minimum "advance at 90° change of heading,"/ and "tactical diameter" **is defined** by the "transfer at 180° change of heading."/

2. Initial turning ability

   The initial turning ability **is defined**/ by the change-of-heading response to a moderate helm/ in terms of heading deviation per unit distance sailed./

3. Course-keeping ability

   Course-keeping ability **is** a measure of the ability of the ship/ to maintain a straight path/ on a predetermined course/ without excessive oscillations of rudder of heading./

4. Yaw-checking ability

   The yaw-checking ability of a ship **is** a measure of the response to counter-rudder/ applied in a certain state of turning./

5. Stopping ability

   Stopping ability **is measured** by the "track reach" and "time to dead in water"/ realized in a stop engine—full astern maneuver/ performed after a steady approach at full test speed./

《語　注》

1. broadly：大まかに
2. evasive：回避するための
3. decelerating：減速
4. advance：縦距
5. tactical diameter：旋回径
6. transfer：横距
7. deviation：逸脱，偏差
8. track reach：到達距離
9. oscillation：振動
10. dead in water：停止

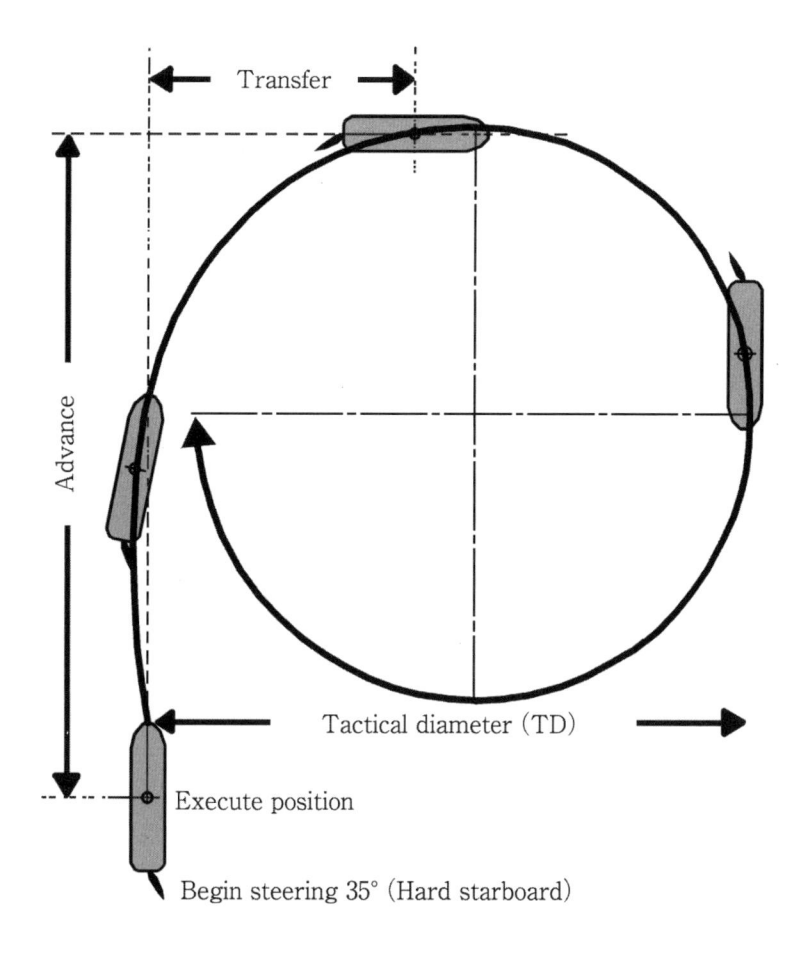

Reference：
International Mariners Management Association of Japan (IMMAJ), 2008, THE BEST SEAMANSHIP —A Guide to Ship Handling—, Chapter 1 Maneuvering Capability of Ships

Unit 43 当直に関する基準

Standards Regarding Watchkeeping

Planning prior to each voyage
Prior to each voyage/ the master of every ship **shall ensure**/ that the intended route from the port of departure to the first port of call **is planned**/ using adequate and appropriate charts and other nautical publications/ necessary for the intended voyage,/ containing accurate, complete and up-to-date information/ regarding those navigational limitations and hazards/ which **are** of a permanent or predictable nature,/ and which **are** relevant to the safe navigation of the ship./

Deviation from planned route
If a decision **is made**,/ during a voyage,/ to change the next port of call of the planned route,/ or if it **is** necessary for the ship to deviate substantially from the planned route for other reasons, /then an amended route **shall be planned**/ prior to deviating substantially from the route originally planned./

Look-out
A proper look-out **shall be maintained** at all times/ in compliance with rule 5 of the International Regulations for Preventing Collisions at Sea, 1972/ and **shall serve** the purpose of:/

.1 maintaining a continuous state of vigilance/ by sight and hearing,/ as well as by all other available means,/ with regard to any significant change in the operating environment;/

.2 fully appraising the situation and the risk of collision,/ stranding and other dangers to navigation;/ and

.3 detecting ships or aircraft in distress, shipwrecked persons, wrecks, debris and other hazards to safe navigation./

In determining that the composition of the navigational watch **is** adequate to ensure/ that a proper lookout **can** continuously **be maintained**,/ the master **shall take into account** all relevant factors,/ including those described in this section of the Code,/ as well as the following factors:/

.1 visibility, state of weather and sea;/

.2 traffic density, and other activities/ occurring in the area/ in which the vessel **is navigating**;/

.3 the attention necessary/ when navigating in or near traffic separation schemes/ or other routing measures;/

.4 the additional workload/ caused by the nature of the ship's functions,/ immediate operating requirements and anticipated manoeuvres;/

.5 the fitness for duty of any crew members on call/ who **are assigned** as members of the watch;/

.6 knowledge of/ and confidence in the professional competence/ of the ship's officers and crew;/

.7 the experience of each officer of the navigational watch,/ and the familiarity of that officer/ with the ship's equipment, procedures, and manoeuvring capability;/

.8 activities taking place on board the ship/ at any particular time,/ including radiocommunication activities/ and the availability of assistance/ to be summoned immediately to the bridge/ when necessary;/

.9 the operational status of bridge instrumentation and controls,/ including alarm systems;/

.10 rudder and propeller control/ and ship manoeuvring characteristics;/

.11 the size of the ship/ and the field of vision/ available from the conning position;/

.12 the configuration of the bridge,/ to the extent such configuration **might inhibit** a member of the watch from detecting by sight/ or hearing any external development;/ and

.13 any other relevant standard, procedure or guidance/ relating to watchkeeping arrangements and fitness for duty/ which **has been adopted** by the Organization./

Taking over the watch

Relieving officers **shall** personally **satisfy** themselves/ regarding the:

.1 standing orders and other special instructions of the master/ relating to navigation of the ship;/
.2 position, course, speed and draught of the ship;/
.3 prevailing and predicted tides, currents, weather, visibility and the effect of these factors upon course and speed;
.4 procedures for the use of main engines to manoeuvre/ when the main engines **are** on bridge control;/ and
.5 navigational situation,/ including but not limited to:/
  .5.1 the operational condition of all navigational and safety equipment/ being used or likely to be used/ during the watch,/
  .5.2 the errors of gyro and magnetic compasses,/
  .5.3 the presence and movement of ships in sight/ or known to be in the vicinity,/
  .5.4 the conditions and hazards likely to be encountered/ during the watch,/ and
  .5.5 the possible effects of heel, trim, water density/ and squat on under keel clearance./

Performing the navigational watch

The officer in charge of the navigational watch **shall**:

.1 **keep** the watch on the bridge;/
.2 in no circumstances **leave** the bridge/ until properly **relieved**;/
.3 **continue** to be responsible for the safe navigation of the ship,/ despite the presence of the master on the bridge,/ until **informed** specifically/ that the master **has assumed** that responsibility/ and (that) this **is** mutually **understood**./

During the watch/ the course steered, position and speed **shall be checked**/ at sufficiently frequent intervals,/ using any available navigational aids necessary,/ to ensure that the ship **follows** the planned course./

The officer in charge of the navigational watch **shall make** regular checks/ to ensure that:

.1 the person steering the ship or the automatic pilot **is steering** the correct course;/
.2 the standard compass error **is determined** at least once a watch/ and, when possible, after any major alteration of course; the standard and gyro-compasses **are** frequently **compared**/ and repeaters **are synchronized with** their master compass;/
.3 the automatic pilot **is tested** manually/ at least once a watch;/
.4 the navigation and signal lights and other navigational equipment **are functioning** properly;/
.5 the radio equipment **is functioning** properly/ in accordance with paragraph 86 of this section;/ and
.6 the UMS controls, alarms and indicators **are functioning** properly./

The officer in charge of the navigational watch **shall use** the radar/ whenever restricted visibility **is encountered** or **expected**, and at all times in congested waters/ having due regard to its limitations./

The officer in charge of the navigational watch **shall ensure**/ that range scales employed **are changed** at sufficiently frequent intervals/ so that echoes **are detected** as early as possible./ It **shall be borne** in mind that small or poor echoes **may escape** detection./

Whenever radar **is** in use, the officer in charge of the navigational watch **shall select** an appropriate range scale/ and **observe** the display carefully,/ and **shall ensure**/ that plotting or systematic analysis **is commenced** in ample time./

《語注》

1. intended route：計画航路
2. predictable nature：予想される自然現象
3. deviation from planned route：予定航路からの離路
4. amended route：変更した航路
5. vigilance：警戒
6. appraise：認識する
7. stranding：座礁
8. traffic separation schemes：分離通航方式
9. other routing measures：他の航路指定方法
10. competence：能力
11. summon：呼び出す
12. configuration：形状
13. prevailing：現在の
14. vicinity：付近
15. squat：沈下
16. under keel clearance：余裕水深
17. mutually：相互に
18. synchronize：同調する
19. in accordance with 〜：〜の規定に定めるところにより
20. congested water：交通輻輳水域
21. have due regard to 〜：〜を十分に考慮する
22. bear in mind：留意する
23. commence：開始する

Reference：
International Maritime Organization (IMO), International Convention on Standards of Training, Certification and Watchkeeping for Seafarers (STCW), 1978, Annex, Part A, Chapter 8

コラム 8

接続詞の that と関係代名詞の that の見分け方

2つの文を結びつける that の用法ですが,どこで見分けるのかが気になるところです。
ヒントは that の後ろにあります。接続詞 that の後ろには完全な文が来ます。たとえば

① I think that he is suitable for the new captain.
（彼は新しい船長としてふさわしいと思います。）

この場合,that は「ということ」のような意味を持ち,文の状況に応じて訳します。
一方,関係代名詞の that の後ろには不完全な文が来ます。

② I will get on board the cruise vessel that was launched last month.
（私は先月進水したクルーズ船に乗ります。）

that の後ろを見ると,いきなり動詞が登場しています。これで主語が欠けているということがわかります。実際には,主語が欠けているのではなく,2つ目の文の主語が関係代名詞の that になったので,足りないように見えているだけなのですが,長い文を読解するときの参考にしてみてください。

Unit 44 航行の安全

Safety of Navigation

Shipborne navigational equipment and systems
All ships/ irrespective of size/ **shall have**:
1. a properly adjusted standard magnetic compass or other means,/ independent of any power supply/ to determine the ship's heading/ and display the reading at the main steering position;/
2. a pelorus or compass bearing device, or other means,/ independent of any power supply/ to take bearings over an arc of the horizon of 360°;/
3. means of correcting heading and bearings/ to true at all times;/
4. nautical charts and nautical publications/ to plan and display the ship's route for the intended voyage/ and to plot and monitor positions/ throughout the voyage./ An electronic chart display and information system (ECDIS) **is** also **accepted**/ as meeting the chart carriage requirements of this subparagraph. Ships/ to which paragraph 2.10 **applies**/ **shall comply** with the carriage requirements for ECDIS detailed therein;/
5. back-up arrangements/ to meet the functional requirements of subparagraph .4,/ if this function **is** partly or fully **fulfilled** by electronic means;/
6. a receiver/ for a global navigation satellite system or a terrestrial radionavigation system, or other means,/ suitable for use at all times/ throughout the intended voyage/ to establish and update the ship's position/ by automatic means;/
7. if less than 150 gross tonnage and if practicable,/ a radar reflector or other means,/ to enable detection by ships navigating by radar at both 9 and 3 GHz;/
8. when the ship's bridge **is** totally **enclosed**/ and unless the Administration **determines** otherwise,/ a sound reception system, or other means,/ to enable the officer in charge of the navigational watch/ to hear sound signals and determine their direction;/
9. a telephone, or other means,/ to communicate heading information/ to the emergency steering position,/ if provided./

Steering gear: testing and drills
Within 12 hours before departure,/ the ship's steering gear **shall be checked** and **tested** by the ship's crew./ The checks and tests **shall include**:
1. the full movement of the rudder/ according to the required capabilities of the steering gear;/
2. a visual inspection of the steering gear and its connecting linkage;/ and
3. the operation of the means of communication/ between the navigation bridge and steering gear compartment./

Nautical charts and nautical publications
Nautical charts and nautical publications, /such as sailing directions, lists of lights, notices to mariners, tide tables and all other nautical publications/ necessary for the intended voyage,/ **shall be** adequate and up to date./

Danger Messages
The master of every ship/ which **meets** with dangerous ice, a dangerous derelict, or any other direct danger to navigation, or a tropical storm,/ or **encounters** sub-freezing air temperatures/ associated with gale force winds/ causing severe ice accretion on superstructures,/ or winds of force 10 or above on the Beaufort scale/ for which no storm warning **has been received**,/ **is bound to communicate** the information/ by all means at his disposal/ to ships in the vicinity, and also to the competent authorities./

Distress Situations: Obligations and procedures

The master of a ship at sea/ which is in a position to be able to provide assistance/ on receiving information from any source that persons are in distress at sea,/ is bound to proceed with all speed to their assistance,/ if possible informing them/ or the search and rescue service that the ship is doing so./ This obligation to provide assistance applies/ regardless of the nationality or status of such persons or the circumstances/ in which they are found. If the ship receiving the distress alert is unable/ or, in the special circumstances of the case,/ considers it unreasonable or unnecessary to proceed to their assistance,/ the master must enter/ in the log-book/ the reason for failing to proceed to the assistance of the persons in distress,/ taking into account the recommendation of the Organization,/ to inform the appropriate search and rescue service accordingly./

Safety navigation and avoidance of dangerous situations

Prior to proceeding to sea,/ the master shall ensure/ that the intended voyage has been planned using the appropriate nautical charts and nautical publications for the area concerned,/ taking into account the guidelines and recommendations developed by the Organization./

The voyage plan shall identify a route/ which:

1. **takes into account** any relevant ships' routeing systems;/
2. **ensures** sufficient sea room for the safe passage of the ship throughout the voyage;/
3. **anticipates** all known navigational hazards and adverse weather conditions;/ and
4. **takes into account** the marine environmental protection measures that apply,/ and avoids, as far as possible, actions and activities which could cause damage to the environment./

《 語 注 》
1. irrespective of：にかかわりなく
2. pelorus：方位盤、ペロルス
3. electronic chart display and information system：電子海図情報表示システム
4. global navigation satellite system：全地球航法衛星システム
5. terrestrial radionavigation system：地上無線航法システム
6. practicable：実施可能な
7. sound reception system：音響受信装置
8. derelict：遺棄物
9. ice accretion：着氷
10. superstructure：上部構造物
11. vicinity：付近
12. competent authority：監督当局

Reference：
International Maritime Organization (IMO), International Convention for the Safety of Life At Sea (SOLAS), 1974, Annex, Chapter 5

Unit 45 　捜索パターン

Search Patterns

Expanding Square Search (SS)
- Most effective/ when the location of the search object **is known**/ within relatively close limits./
- The commence search point **is** always the datum position./
- Often appropriate for vessels or small boats to use/ when searching for persons in the water/ or other search objects with little or no leeway./
- Due to the small area involved,/ this procedure **must not be used** simultaneously/ by multiple aircraft at similar altitudes/ or by multiple vessels./
- Accurate navigation **is required**;/ the first leg **is** usually **oriented** directly into the wind/ to minimize navigational errors./
- It **is** difficult for fixed-wind aircraft to fly legs close to datum/ if S **is** less than 2 NM./

Track Line Search (TS)
- Normally **used**/ when an aircraft or vessel **has disappeared**/ without a trace along a known route./
- Often **used**/ as initial search effort/ due to ease of planning and implementation./
- **Consists** of rapid and reasonably thorough search/ along intended route of the distressed craft./
- Search **may be** along one side of the track line/ and **return** in the opposite direction on the other side (TSR)./
- Search **may be** along the intended track and once on each side,/ then search facility **continues** on its way and **does not return** (TSN)./
- Aircraft **are** frequently **used** for TS/ due to their high speed./
- Aircraft search height usually 300 m to 600 m (1000 ft. to 3000 ft.) during daylight or 600 m to 900 m (2000 ft. to 3000 ft.) at night./

Parallel Track Search (PS)
- **Used** to search a large area/ when survivor location **is** uncertain./
- Most effective over water or flat terrain./
- Usually **used**/ when a large search area **must be divided** into sub-areas for assignment/ to individual search facilities on-scene at the same time./
- The commence search point **is** in one corner of the sub-area,/ one-half track space inside the rectangle from each of the two sided forming the corner./
- Search legs **are** parallel to each other/ and to the long sides of the sub-area./

《 語　注 》
1. expanding square search：拡大方形捜索
2. commence search point：捜索開始地点
3. simultaneously：同時に
4. track line search：航路線捜索
5. implementation：履行
6. parallel track search：平行トラック捜索
7. assignment：割り当てられた仕事、分担
8. rectangle：長方形

Reference：
International Maritime Organization (IMO) and the International Civil Aviation Organization (ICAO), International Aeronautical and Maritime Search and Rescue (IAMSAR) Manual, Volume 3, Mobile Facilities, Section 3

Unit 46　船位通報制度

Ship Reporting System

The ship reporting system **should provide** up-to-date information on the movements of vessels/ in order, in the event of a distress incident,/ to:
1. reduce the interval between the loss of contact with a vessel and the initiation of search/ and rescue operations/ in cases where no distress signal **has been received**;/
2. permit rapid identification of vessels/ which **may be called upon** to provide assistance;/
3. permit delineation of a search area of limited size/ in case the position of a person, a vessel or other craft in distress **is unknown** or **uncertain**;/ and
4. facilitate the provision of urgent medical assistance or advice./

Ship reporting systems **should satisfy** the following requirements:/
1. provision of information,/ including sailing plans and position reports,/ which **would make** it possible to determine the current and future positions of participating vessels;/
2. maintenance of a shipping plot;/
3. receipt of reports/ at appropriate intervals/ from participating vessels;/
4. simplicity in system design and operation;/ and
5. use/ of internationally agreed standard ship reporting format and procedures./

A ship reporting system **should incorporate** the following types of ship reports/ in accordance with the recommendations of the Organization:/
1. Sailing plan;/
2. Position report;/ and
3. Final report./

《 語　注 》
delineation：画定

Reference：
International Maritime Organization (IMO), International Convention on Maritime Search and Rescue (SAR Convention), 1979, Annex, Chapter 5

コラム 9

関係代名詞 that の使い方

関係代名詞 that の働きが 2 つの文をつなげるということはわかるけど，どう訳していいのか悩んでいる人もいるかもしれません。ここでは，どのようにくっついて，どう訳せばいいのか，簡単に説明します。

まずは，わかりやすく真ん中で半分に分かれる文です。

① The Suez Canal **is** a famous canal that **took** ten years to build.
　　（スエズ運河はつくるのに 10 年かかった有名な運河です。）

この文は that で半分に分かれています。that を中心に動詞が両方に分かれているのがわかるでしょう。
少し難しいと思われるのが，次の文のタイプです。

② Ships that **use** Panama Canal **pay** dues.
　　（パナマ運河を使う船舶は料金を支払います。）

こっちのタイプは，that の後ろに動詞が 2 つともあるので，どこで分ければいいのか迷ってしまうこともあるようです。でも，動詞を目印にして，2 つ目の動詞の前で区切ることを覚えれば，訳しやすくなるでしょう。

　　Ships that **use** Panama Canal/ **pay** dues.
　　（パナマ運河を使う船舶は / 料金を支払います。）

Unit 47 運航に伴う油の排出規制

Control of Operational Discharge of Oil

Requirements for machinery spaces of all ships
Control of discharge of oil

Subject to the provisions of regulation 4 of this annex and paragraphs 2, 3, and 6 of this regulation,/ any discharge into the sea of oil or oily mixtures from ships **shall be prohibited**./

Discharges outside special areas

Any discharge into the sea of oil or oily mixtures from ships of 400 gross tonnage and above **shall be prohibited**/ except when all the following conditions **are satisfied**:/

1. the ship **is proceeding** en route;/
2. the oily mixture **is processed** through an oil filtering equipment/ meeting the requirements of regulation 14 of this Annex;/
3. the oil content of the effluent without dilution **does not exceed** 15 parts per million;/
4. the oily mixture **does not originate** from cargo pump-room bilges on oil tankers;/ and
5. the oily mixture, in case of oil tankers, **is not mixed** with oil cargo residues./

Requirements for the cargo area of oil tankers
Control of discharge of oil
Discharges outside special areas

Subject to the provisions of regulation 4 of this Annex and paragraph 2 of this regulation,/ any discharge into the sea of oil or oily mixtures from the cargo area of an oil tanker **shall be prohibited**/ except when all the following conditions **are satisfied**:/

1. the tanker **is not** within a special area;/
2. the tanker **is** more than 50 nautical miles from the nearest land;/
3. the tanker **is proceeding** en route;/
4. the instantaneous rate of discharge of oil content **does not exceed** 30 litres per nautical mile;/
5. the total quantity of oil discharged into the sea **does not exceed** [for tankers delivered on or before 31 December 1979, as defined in regulation 1.28.1], 1/15,000 of the total quantity of the particular cargo/ of which the residue **formed** a part,/ and [for tankers delivered after 31 December 1979, as defined in regulation 1.28.2], 1/30,000 of the total quantity of the particular cargo/ of which the residue **formed** a part;/ and
6. the tanker **has**/ in operation/ an oil discharge monitoring and control system and a slop tank arrangement/ as required by regulations 29 and 31 of this Annex./

《 語 注 》

1. oily mixture：油性混合物
2. gross tonnage：総トン数
3. oil filtering equipment：油除去装置
4. oil content：油分
5. effluent：廃水
6. dilution：希釈
7. cargo pump-room：貨物油ポンプ室
8. oil cargo residue：貨物油残留物
9. special area：特別海域
10. instantaneous rate of discharge：瞬間排出率
11. total quantity：総量
12. slop tank：スロップタンク

Reference：
International Maritime Organization (IMO), International Convention for the Prevention of Pollution from Ships (MARPOL), 1973, Annex 1, Chapter 3, 4

コラム10

so ～ that と so that の使い方

so ～ that… を「とても～なので…」と訳すことはかなりの頻度で登場しているので，理解している人も多いことでしょう。

① The waves were so high that the ship could not move forward smoothly.
　　（波がとても高かったので，船はスムーズに前に進めなかった。）

また，so that という so と that の間に何も単語が入らない文を見ることがあるかもしれません。ここでは，その2つの違いについて触れておきます。

①の文では，that の後ろには「結果」を表す文が続きます。
これに対して，so that は次のような文で現れます。

② A ferry is designed so that it can carry both passengers and vehicles.
　　（フェリーは乗客と乗り物の両方を運ぶように設計される。）

②の文では，so that の後ろが目的を表す文になります。
なので，so と that の間に何か来れば前から訳していけばいいし，so と that の間に何も入っていなければ，後ろの方から「～するために」と訳せばいいのです。

CHAPTER 4

専門英語 機関

Unit 48　調整の許容範囲

Tolerance Limit of Alignment

Alignment must be made within 5/100 mm (7/100 to 8/100 mm may be acceptable)./
Adjust the clearance/ between the mating faces of a motor and pump couplings/ by 3 to 4 mm./ Basically, adjustment shall be made on the motor side./ Measurement shall be done vertical/ and sideway difference shall be 5/100 mm or less./

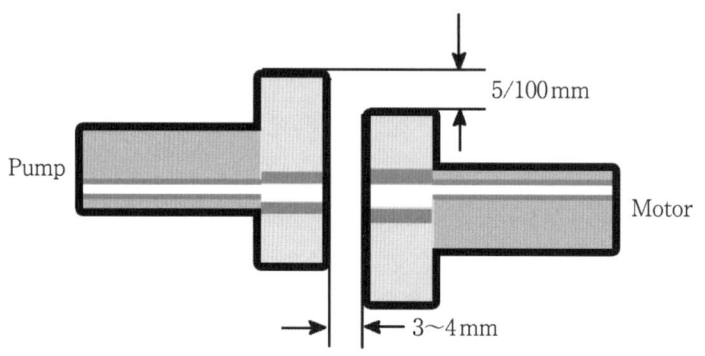

(1) **Adjust** the clearance/ between the mating faces of the motor and pump couplings/ on upper and lower sides (measuring instrument: fleer gauge)./

For adjusting with a liner,/ **insert** the liner/ on the front or rear side/ of the mating faces./ **Use** a liner of the same thickness/ on both left and right sides./ **Turn** the coupling/ by 180 deg. to measure the clearance./

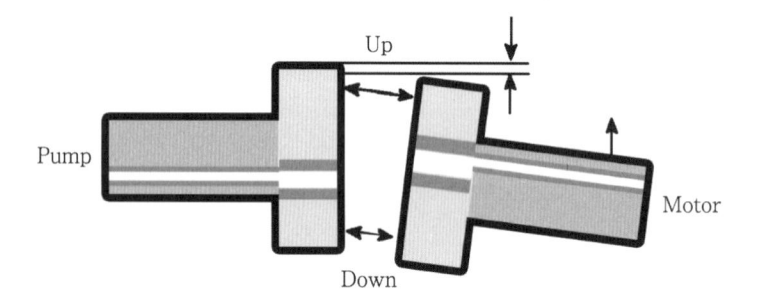

(2) Adjustment of the height (measuring instrument: dial gauge)
(It is a) Method to mount a dial gauge/ corresponding to the shape of coupling and other factors./
In adjusting with liners,/ **insert** liners of the same thickness/ at 4 positions./

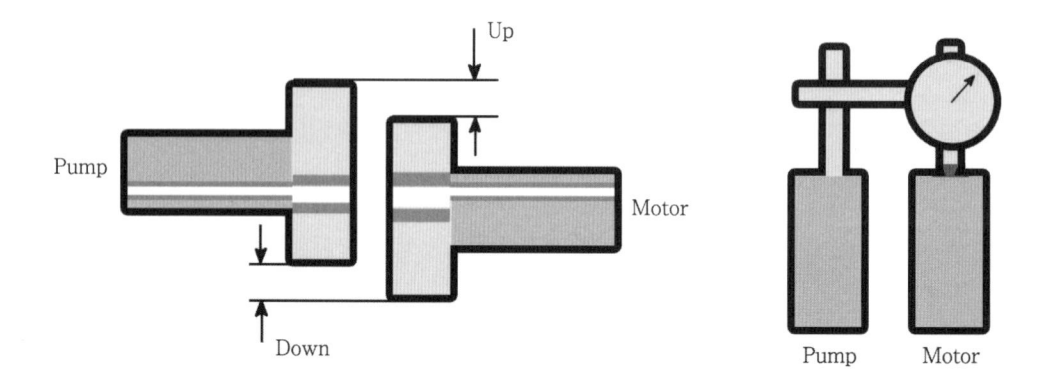

(3) Adjustment of a miss-alignment and a clearance/ between mating faces of the motor and pump couplings./ First **adjust** a lateral miss-alignment,/ then **adjust** a clearance/ between the mating faces of the motor and pump couplings/ by slightly tightening the anchor bolts of the motor./ **Repeat** these adjustments./

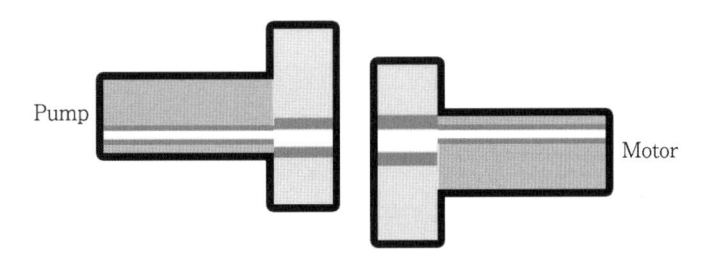

Pump　　　　Motor

(4) **Repeat** procedures of the above (3)./
(5) After finishing the adjustment,/ **tighten** anchor bolts./

《 語　注 》
1. tolerance：(直訳：許容限界) ばらつきが許される限界値
2. alignment：位置合わせ
3. adjust：調整
4. clearance：すき間
5. mating faces：接触面
6. front or rear side：前後面
7. dial gauge：ダイヤルゲージ
8. mount：〜を取り付ける、用意する

Reference：
International Mariners Management Association of Japan (IMMAJ), 2008, THE BEST SEAMANSHIP ―A Guide to Engine Skills―, Chapter 6 Tolerance Limit of Alignment

Unit 49 非破壊試験

Non-Destructive Test (Test without Destroying the Material)

Detecting Test by Penetrant

(1) Process

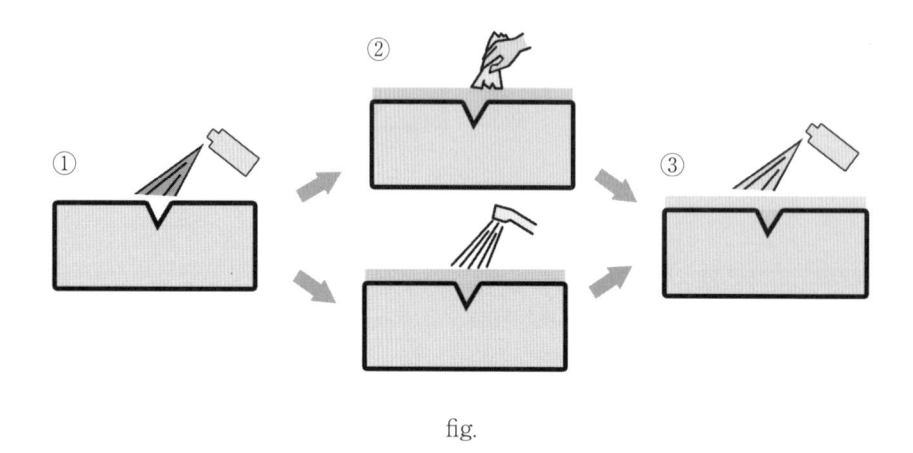

fig.

(2) Features

This test is applicable to either metal or nonmetal materials/ unless they are affected by the penetrating liquid/ or have a porous nature./

This test is applicable/ against any extended direction of the defects./

This test is useful for materials/ which surface are damaged and extended interior/ as a cave./

(3) Implementation of test
- **Clean up** the surface/ for removing dust and oil./
- **Apply** the penetrating liquid (Process fig.-①)
    Fluorescent penetrant: to observe through ultra violet rays in the dark
    Dying penetrant: color vision in the light
- **Clean up** the surface (Process fig.-②)
    Wipe out the surface liquid/ except the one interior of the material./
    When the applying penetrant includes the additive of an emulsifier,/ the surface should be washed by water./

(4) Development: formation of indicating defects pattern (Process fig.-③)

Dry developing method:	Dry developing solution is used
Quick dry developing method:	developing solution/ with a voltaic solvent is used
Wet developing method:	Developing solution/ dispersed in water is used
Non development method:	process by heart treatment/ instead of developing solution

(5) Finish the test

Remove the remaining penetrant by wiping out and/or flushing by water./

(1)

(2)

(3)

(4)

(5)

(6)

(7)

(8)

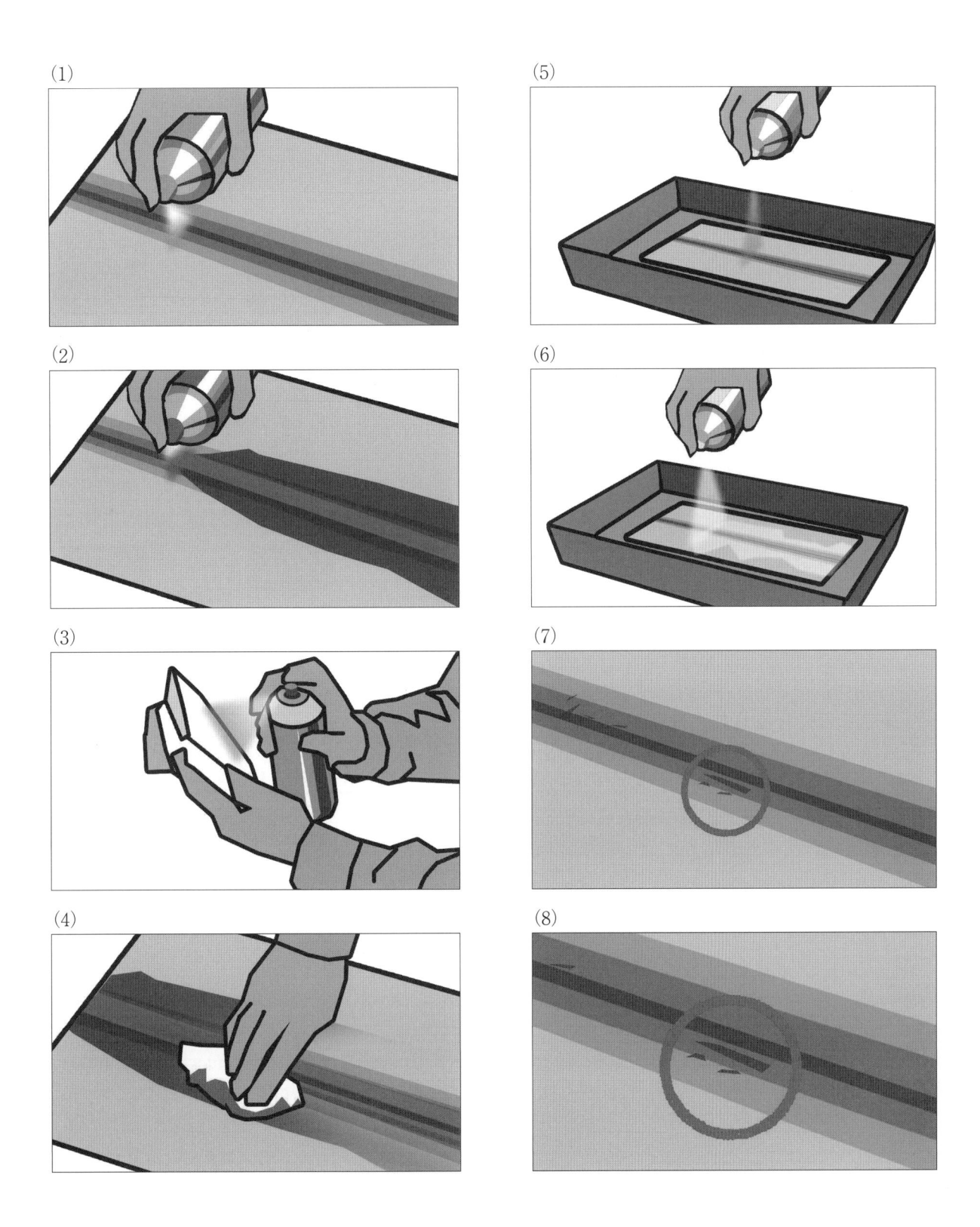

《 語 注 》

1. Non-Destructive Test：非破壊試験
2. Detecting Test：検出試験
3. penetrant：浸透剤
4. applicable：適用できる
5. metal or nonmetal materials：金属か非金属材料
6. porous nature：(直訳) 孔性

Reference：

International Mariners Management Association of Japan (IMMAJ), 2008, THE BEST SEAMANSHIP ―A Guide to Engine Skills―, Chapter 9 Non-Destructive Test (Test without Destroying the Material)

Unit 50 ダイオード

Diode

Use of diode

A diode is a semiconductor device/ that works as a check valve in electrical circuits./ It is also called rectifier./ Check valves prevent water, steam, or fuel in the piping/ from flowing backward/ and are used to let the liquid flow/ in one direction only./ Diodes allows current/ to flow only in one direction/ and do not allow it to flow/ in the opposite direction./ Positive or negative signal is acquired/ using this function of diode./ Diode also converts alternate current/ into direct current./

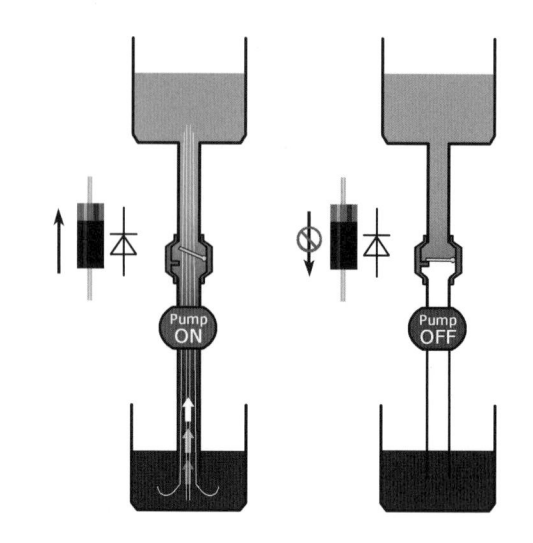

Structure of diodes

As shown in Fig. 2-8 (a),/ a diode is composed of a combination/ of P type semiconductors and N type semiconductors./ It is indicated in electrical circuits/ as shown in Fig. 2-8 (b)./ The terminal on the P type semiconductor side is called an anode,/ and the terminal on the N type semiconductor side is called a cathode./ The junction surface becomes electrically neutralized/ as the electron hole in the P type semiconductor and a free electron in the N type semiconductor pull/ against each other./
This is called a depletion layer./

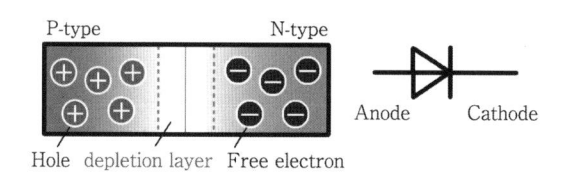

The DC voltage source is connected with the semiconductor of this P-N junction/ as shown in Fig. 2-9./ At this time,/ the hole in the P type semiconductor is attracted/ towards the negative side of the DC voltage source./ Moreover,/ a free electron of the N type semiconductor is drawn/ to the positive side./ Therefore,/ the depletion layer in the P-N junction expands,/ thus disturbing the movement of the electric charge./ As a result,/ the current does not flow/ through the diode./

Meanwhile,/ the state of connecting the DC power supply/ in the opposite direction is shown/ in Fig. 2-10./ As mentioned earlier,/ the hole in the P type semiconductor is attracted/ towards the negative side/ and a free electron of the N type semiconductor is attracted/ towards the positive side./ As a result,/ carriers from both sides jump into the opposite semiconductors/ and the current starts flowing./

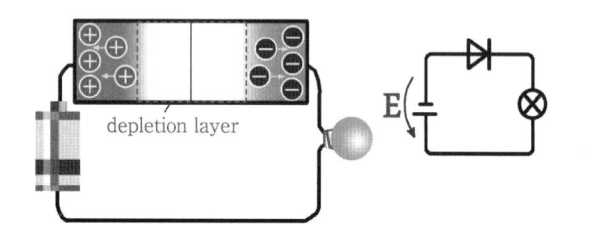

Fig. 2-8 (a)   Fig. 2-8 (b)

Fig. 2-9 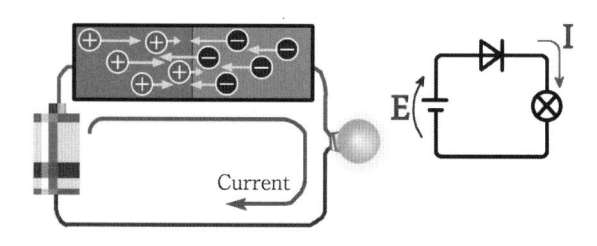   Fig. 2-10

Types of diode

There **are** various types of diode. Some examples **are listed**/ below (Figs. 2-11)./

(a) Diode rectifier

It **converts** single-phase AC voltage or three-phase AC voltage/ into DC current./
It **is composed** of 4 or 6 diodes/ and **has** a plastic (synthetic resin) cover./

(b) High-speed recovery diode

It **is used**/ when a high-speed operation/ such as DC power supply/ or inverter functionality **is needed**.

(c) Zener diode

It **is used**/ in a circuit that **supplies** constant voltage./ It **simplifies** the constant voltage circuit,/ but **decreases** efficiency./

(d) LED (Light emitting diode)

The P-N junction **emits** light/ when current **is passed** through it./ Its main advantages **are** low power consumption and a long life./

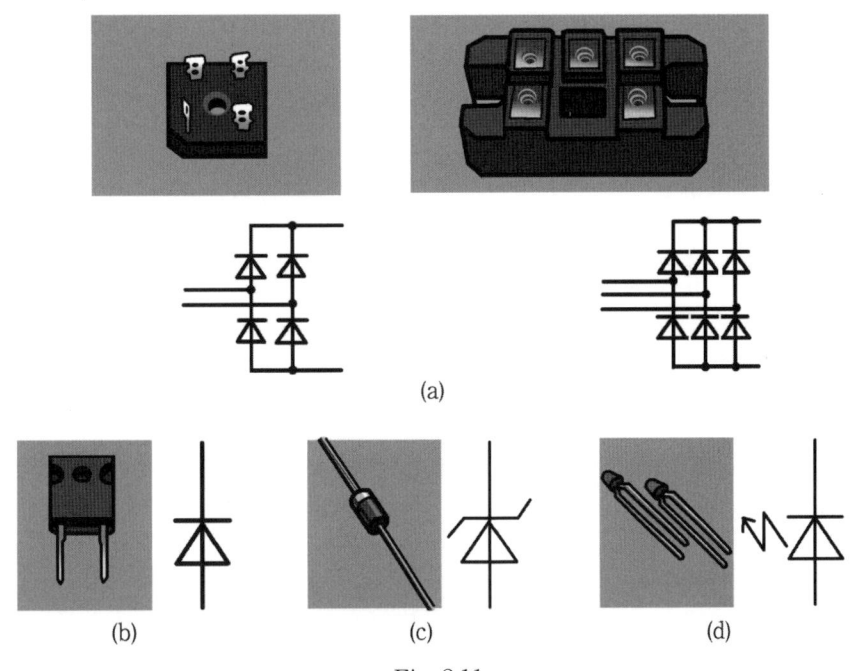

Fig. 2-11

《 語 注 》

1. semiconductor device：半導体
2. check valve：逆止弁
3. rectifier：整流器
4. one direction only：一方向のみ
5. positive or negative signal：正か負の信号
6. convert：変換
7. alternate current：交流
8. direct current：直流

Reference：

International Mariners Management Association of Japan (IMMAJ), 2008, THE BEST SEAMANSHIP —Fundamentals of Maritime Electronic Apparatus—, Chapter 2 Diode

Unit 51　ターボチャージャーのサージング

Main Engine Turbocharger Surging

1. Main engine type
Sulzer 7RTA84M (Turbocharger type: MET83SC × 2 sets)

2. Process of Breakdown
After departing from Hong Kong,/ while increasing to programmed ocean cruising speed,/ when the main engine speed reached the 71 rpm (52% load) range,/ surging occurred from the main engine No. 2 Turbocharger./

After speed increase was stopped,/ various countermeasures were instituted/ in the form of Turbocharger casing washing, blower washing etc.,/ but to no effect./ Since the exhaust temperature could be lowered/ by using the auxiliary blower in parallel,/ it was decided/ that operation would be continued/ in this condition./ However, even after this,/ gradual increase of exhaust gas temperature continued./ As the alarm temperature (480°C) was again attained,/ we were forced to reduce speed./ Consequently,/ a Turbocharger inlet temperature limit of 550°C was set,/ the speed was increased/ as far as operation would allow./ Although every effort was made to sustain the above conditions,/ as the rise in exhaust gas temperature would not stop,/ an output of 49 rpm was the limit./ A switch over to A oil was carried out/ and the vessel entered Singapore port./

3. Outline of breakdown and Repair
At Singapore port,/ overhaul of the No. 1 and 2 Turbochargers was carried out. When the exhaust manifold and manhole were inspected,/ about 70% of the Turbocharger inlet grid/ inside the No. 1 exhaust manifold/ and about 75% of the Turbocharger inlet grid/ inside the No. 2 exhaust manifold were blocked up/ with adhesive solid products./ In addition,/ as internal fouling of the exhaust gas economizer was severe,/ washing was carried out/ at the same time./

Rotor removal and cleaning were carried out/ on both Turbochargers,/ fouling of the No. 1 Turbocharger blower, diffuser, No. 2 Turbocharger nozzle ring/ and turbine blade was discovered./

Arrangements were made for an offshore repair company/ to accomplish the series of activities./

4. Causes
When the fuel used by this vessel was analyzed (loaded at LA, IF380,)/ the general property values exceeded a density of 0.991 g/cm³, ash content was 6~7 times greater/ than ordinary fuel, residual carbon content, heptane insoluble content, and dry sludge amount exceeding 0.1 wt% and (ash content) was detected in large amounts./ Judging from these facts,/ it was made known/ that the fuel was heavy crude/ and contained large amount of impurities.

Again, from metallic analysis tests,/ FCC catalytic particles ($AL_2 O_3 SiO_2$) and sediment were detected in large amounts./ Moreover, Na and Ca content was abnormally great./ This would make it easy/ for problems to occur/ relating to large sludge quantity expulsion/ from FO purifier and filter blockage trouble,/ as well as fouling of the combustion chamber and exhaust system, fouling of the turbocharger,/ and (this) increased the possibility/ for surging trouble to result at the same time./

For the above reasons,/ although sub-standard fuel oil properties caused the breakdown this time,/ as crack welding work for the FO double bottom tank was part of the dock entry repair work/ accomplished after completing voyage,/ dragging of the tank bottom was carried out for this vessel./ Hence, the trouble was caused/ by the intermixing of these sub-standard impurities/ accumulated at the tank bottom./

(Fuel oil test results table)

Density	g/cm³	0.9918
Kinematic viscosity	Cst	356
Ignition point	°C	105
Water content	vol %	0.40
Residual carbon	wt %	15.0
Ash content	wt %	0.28
Sulfur content	wt %	1.54
Pentane insolubles	wt %	13.5
Toluene insolubles	wt %	0.18
Dry sludge (SHFT)	wt %	0.1095
FCC catalysts		present
Spot test		1-1-1
CCAI		853.4
Metal analysis SiO_2	ppm	289.9
Na	ppm	389.0
V	ppm	29.4
Al	ppm	11.9
Ca	ppm	538.2

《語　注》

1. after departing：出航後
2. surging：サージング
3. countermeasures：対策
4. A oil：A重油
5. overhaul：分解点検（修理）
6. be inspected：調査される
7. manifold：（内燃機関の）多岐管
8. adhesive：粘着性のある
9. solid product：固体物
10. fouling：付着物
11. diffuser：拡散器
12. sediment：沈殿物

Reference：
International Mariners Management Association of Japan (IMMAJ), 2008, THE BEST SEAMANSHIP —Marine Engine Trouble Case Ⅱ—, Chapter 1 Marine Diesel Engine (Ⅱ)

Unit 52 ピストンクラウンの損傷（バナジウムアタック）

Piston Crown Top Burn Damage

Ship type	VLCC
Date of build	April 1993 NKK Corporation, Tsu factory
Navigation route	Japan ~ PG
Trouble machinery	
Category	Main engine
Name	Piston crown
Type	SULZER 7RTA-84M
Maker	Diesel United
Date of trouble	February 12, 1998

1. Breakdown outline
After departure from PG loading port during S/B operation,/ an M/E No. 4 cylinder crankcase oil mist high density alarm **occurred**./

After moving to a safe ocean area,/ each part of the main engine **was inspected**./ A hole **was confirmed**/ on the top of the No. 4 cylinder piston crown/ where cooling oil **was leaking**./ The alarm condition **occurred**/ as combustion gas **entered** into the damaged area/ during operation/ and finally **intruded** into the crankcase/ via the cooling oil return line./

2. Causes
The surface of the detached piston crown **was pock marked** (refer to the attached photograph)./ Damage to areas/ directly hit by fuel spray **was** especially severe./ The broken hole **was** also **located**/ in this part at dimensions of 13 mm × 5.2 mm./ An inspection of the piston crowns of other cylinders **revealed** similar surface burn damage/ in pock marked configuration,/ although the thickness of the metal **remained**/ within the range of the manufacturer's permissible thickness standards./

The condition of the broken hole closely **resembles** the case of piston top broken hole/ indicated in the NK (NIPPON KAIJI KYOKAI) journal, issue No. 240 (summary of 1995 fiscal year engine damage)./ Therefore, it **is surmised** that the broken hole **was caused**/ by high temperature corrosion (vanadium attack)./

From the time (when) this vessel **entered** service,/ black lacquer **had started** forming on the cylinder liner/ and we **have had** a hard time devising countermeasures./ As one measure,/ on October 1995,/ the fuel valve **was switched** to an improved model./ Compared to the former model,/ the hole angle of the improved model **was** slightly **tilted**/ toward the inside/ to reduce the amount of spray/ directly hitting the liner wall./

As a result of using this improved fuel valve,/ although black lacquer **was suppressed**,/ there **was** an increase/ in the amount of fuel/ reaching the top of the piston crown./ Moreover, the temperature at the crown top **increased** further than before/ to a point which **made** high temperature corrosion (be) more likely to occur./

3. Repair
Replacement with a spare piston./

4. Countermeasures

Signs of high temperature corrosion **are** apparent/ on all cylinders./ Furthermore,/ as corrosion **is not occurring** on same model engines/ which **employ** the former type of fuel valve nozzle,/ there **is** a need/ to replace these engines/ with a fuel valve nozzle/ possessing an appropriate hole angle./

Furthermore,/ the manufacturer **has indicated** the following countermeasures/ within the servicing information./
1) Ensuring good combustion via sufficient fuel pre-treatment
2) Usage of a fuel valve with good atomization
3) Avoiding torque-rich operation
4) Removal of scaling and carbon adhesion to the wall surface of the piston crown oil chamber and cooling hole.

In addition to these countermeasures,/ corrosion-resistance **increases**/ when the piston crown surface **is treated** with inconel filling./

《語 注》
1. was leaking：漏れていた
2. via：経由して
3. pock-marked：小さな穴状に（きず）あとがついて
4. configuration：形状、輪郭、外形
5. permissible：許される程度の（cf. permit）
6. black lacquer：ラッカー（黒）
7. measures：対策，手段
8. countermeasure：対応策
9. be tilted：傾けられる
10. high temperature corrosion：高温腐食
11. sign of：〜の兆候
12. corrosion-resistance：耐食性

Reference：
International Mariners Management Association of Japan (IMMAJ), 2008, THE BEST SEAMANSHIP —Marine Engine Trouble Case Ⅰ—, Chapter 1 Marine Diesel Engine (Ⅰ)

Unit 53 ピストンクラウンの焼損

Main Engine Piston Crown Fire Side Burn Damage

Ship type	Bulker
Date of build	October 1994
Navigation route	Tramper
Trouble machinery	
Category	Main engine
Name	Piston
Type	Mitsubishi SULZER 6RTA62
Maker	Mitsubishi Heavy Industries, Ltd.
Date of trouble	January 30, 1999 during anchorage

1. Breakdown outline

During anchorage,/ as the main engine No. 2 cylinder **had exceeded** 9000 hours of usage time,/ the piston **was drawn** for regular inspection and maintenance./ It **was discovered**/ that the fire side of the piston crown **had incurred** burn damage of 25 mm/ which substantially **exceeds** the manufacturer's usage standard of 10 mm./

When other cylinders **were inspected** from the scavenging port,/ it **was discovered**/ that they also **had incurred** similar burns and abrasion./ As there **was** only one spare piston crown **available**,/ arrangements **were made** for the immediate delivery of piston crown replacement parts./ In view of the anchorage time and delivery time for spare piston crowns,/ the No. 6 cylinder piston/ which **had incurred** burns and wearing damage to an extent second after the No. 2 cylinder/ **was drawn** and **replaced** with the spare crown./

2. Probable causes

Abrasion damage to the piston crown fire side generally **occurs**/ at the part of direct exposure to fuel valve spray./ In this instance as well,/ it **is assumed**/ that abnormal wearing out and expansion of the fuel valve nozzle hole diameter **was** the main cause./

Abnormal wearing out and expansion of the fuel valve nozzle hole diameter **is surmised** to be linked with the use of poor-quality fuel. Furthermore,/ it **was** also **surmised** that there **were** shortcomings in the inspection and maintenance of the fuel valve./

3. Repair

Fuel valve nozzles/ which **were** abnormally **worn** at their nozzle holes/ **were replaced**/ and arrangements **were made** for the delivery of spare nozzles/ in the new year./

As for the remaining piston crowns with burns and abrasion damage,/ they **were** successively **replaced**/ when the opportunity **arose**./

4. Countermeasures

As the possibility for poor-quality fuel to be loaded in the future **is** high,/ the following measures **shall be implemented**:
 1) To shorten the regular inspection and maintenance intervals/ for fuel valves./
 2) To pay attention to the changes/ in exhaust temperatures/ of each cylinder./
 3) To operate/ by narrowing down the flow of liquid/ that **is passed**/ through the fuel purifier./

4) To conduct thorough inspections/ of the piston ring,/ liner, piston crown fire side etc./ via the scavenging port/ when opportunity **allows**.

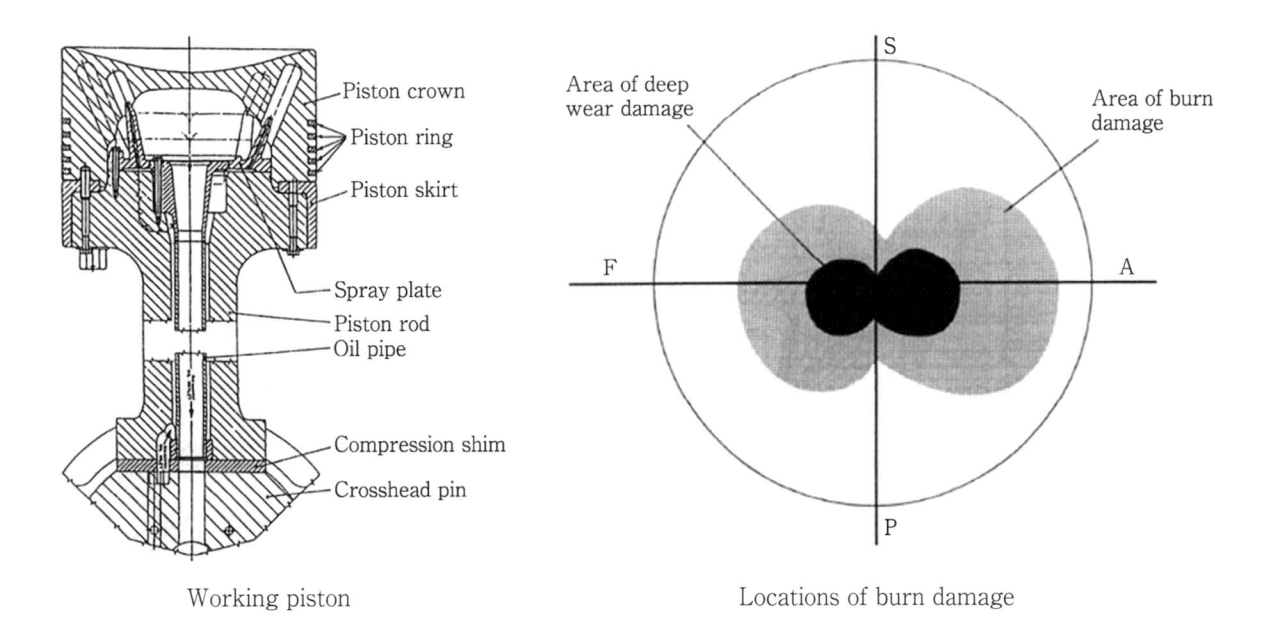

Working piston

Locations of burn damage

Piston crown fire side burn damage diagram

《 語 注 》

1. during anchorage：停泊中
2. had exceeded：超える
3. usage time：使用時間
4. regular inspection：定期検査
5. incur：（損害などを）被る
6. substantially：大幅に
7. arrangement：手配
8. replacement part：交換部品
9. It was surmised that S + V：that 以下が推測された（It は訳さない。that 以下が主語のように訳す）

Reference：
International Mariners Management Association of Japan (IMMAJ), 2008, THE BEST SEAMANSHIP —Marine Engine Trouble Case Ⅰ—, Chapter 1 Marine Diesel Engine (Ⅰ)

Unit 54 船尾管軸受の損傷

Damage to the Stern Tube Bearing

Ship type	Bulker
Category	Shafting
Name	Stern tube bearing
Type	force-feed lubrication
Date of trouble	July 2000

1. Breakdown outline

As a preventive maintenance system is **employed** for propeller shaft inspection of this vessel,/ LO analysis of the stern tube lubrication oil/ including metal analysis of Fe (iron), Cu (copper), and Sn (tin)/ which **are** metallic constituents of the shaft bearing/ **was carried out** every 6 months./ According to analysis results,/ Sn **began** to be detected/ approximately 1 and a half years/ after the vessel delivery for service operation,/ after which the Sn content **indicated** an increasing inclination./ Although ferrography values for the total amount of metallic abrasion **did not indicate** a marked increase/ 4 months before docking,/ Sn content **reached** 21 ppm./ As the classification society's control criterion (upper limit) of 20 ppm **was** slightly **exceeded**,/ the classification society **recommended** that the interval for LO analysis **be shortened**/ to 3 months./

As the Sn content in sample oil/ immediately before docking **had reached** 27 ppm, the confirmation process for stern tube bearing wear down **was sped up** for this docking./ Measurement results **confirmed** a 0.3 mm drop of the shaft bearing at BOTTOM/ and a large quantity of 2 mm-square metallic fragments **was discovered**/ at the bottom of the stern tube LO drain tank.

From these conditions,/ as damage to the stern tube bearing **was anticipated**,/ the propeller shaft **was removed**/ and the stern tube bearing **was examined**./ Exfoliation of bearing metal at a width of 200~300 mm **was observed**/ at the rear end of the bearing/ extending from the starboard lubrication groove/ to the port side/ at an 8 o'clock angle./ There **was** no choice/ but to remove the stern tube AFT BUSH/ and perform additional bearing re-metalling work./

2. Breakdown causes

The bearing metal damage **was not**/ owing to a problem with shaft alignment or vibration,/ but **was** due to insufficient adhesion of the bearing metal/ with the back metal./

As for the adhesion insufficiency,/ further investigation **ascertained**/ that substrate processing/ during bearing white metal casting **was** unsatisfactory/ and that the bearing manufacturer's process management and quality management **were** insufficient./ It **was decided**/ that the bearing manufacturer **would** hereafter **reevaluate** the casting process and **carry out** post-casting analysis/ of the cross sectional structure/ of test pieces/ as countermeasures./

3. Future countermeasures

If an increasing tendency of the content of Cu and Sn/ which **are** primary metallic constituents in bearing metal/ **is observed**,/ the potential for bearing metal abrasion and damage **will be** high./ Therefore,/ cleaning and inspection of the LO strainer and drain tank among other areas **should be carried out**. It **is** important to check for metal fragment deposition,/ and depending on circumstances,/ to consider moving up the docking date.

Furthermore,/ in the event of stern tube bearing damage,/ as the possibility **is** extremely high for a vessel to be forced out-of-service/ over the long-term,/ for vessels which **do not install** a preventive maintenance system for propeller shaft inspection,/ it **is** desirable that periodic LO analysis/ which **includes** an analysis of metallic content/ **be carried out**.

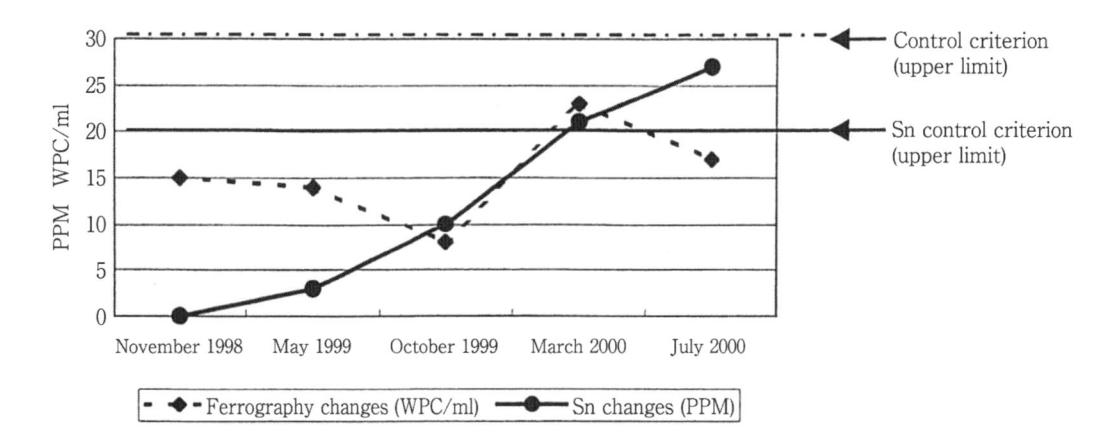

Fig. 1　Trends in stern tube LO analysis results for this vessel

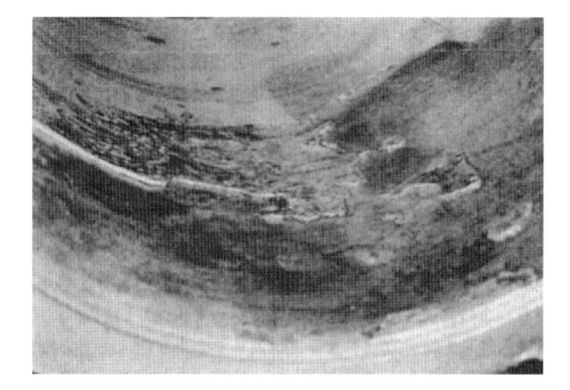

Fig. 2　Damaged area of the stern tube bearing AFT BUSH

《 語　注 》

1. preventive maintenance system：予防保全システム
2. metal analysis：金属検査
3. constituent：成分
4. inclination：傾向
5. ferrography：フェログラフィ（油中の摩耗粒子を磁石などで捕集し，発生部の摩耗量や状況を診断する）
6. for stern tube bearing to wear down：船尾管軸受が摩損する
7. there is no choice but to ～（動詞の原形）：～する以外に選択の余地がない，～するしかない
8. was not owing to ～ but was due to …：～のせいではなくて，…のせいである
9. cross sectional：断面積の

Reference：
International Mariners Management Association of Japan (IMMAJ), 2008, THE BEST SEAMANSHIP —Marine Engine Trouble Case Ⅱ—, Chapter 3 Marine Diesel Engine (Ⅱ)

<編者紹介>
商船高専キャリア教育研究会
商船学科学生のより良きキャリアデザインを構想・研究することを目的に、2007年に結成。
富山・鳥羽・弓削・広島・大島の各商船高専に所属する教員有志が会員となって活動している。

ISBN978-4-303-23347-1

マリタイムカレッジシリーズ
1・2級海技士 はじめての英語指南書

| 2015年8月20日　初版発行 | ⓒ 2015 |
| 2023年4月10日　2版2刷発行 | |

編　者　商船高専キャリア教育研究会　　　　　　　　　検印省略
発行者　岡田雄希
発行所　海文堂出版株式会社
　　　　本社　東京都文京区水道2-5-4（〒112-0005）
　　　　　　　電話 03(3815)3291(代)　FAX 03(3815)3953
　　　　　　　http://www.kaibundo.jp/
　　　　支社　神戸市中央区元町通3-5-10（〒650-0022）
日本書籍出版協会会員・工学書協会会員・自然科学書協会会員

PRINTED IN JAPAN　　　　　　　　　　　印刷　東光整版印刷／製本　誠製本

JCOPY <出版者著作権管理機構 委託出版物>
本書の無断複製は著作権法上での例外を除き禁じられています。複製される場合は，そのつど事前に，出版者著作権管理機構（電話03-5244-5088, FAX 03-5244-5089, e-mail: info@jcopy.or.jp）の許諾を得てください。